Raven Press Series in Physiology
William F. Ganong, M.D., Series Editor

An Introduction to Membrane Transport and Bioelectricity,
Second Edition (1994)
John H. Byrne and Stanley G. Schultz

Respiratory Physiology, Third Edition (1993)
Allan H. Mines

Basic Medical Endocrinology
Second Edition (1994)
H. Maurice Goodman

Cardiovascular Physiology (1988)
Jon Goerke and Allan H. Mines

An Introduction to Membrane Transport and Bioelectricity

Foundations of General Physiology and Electrochemical Signaling

Second Edition

John H. Byrne, Ph.D.

Professor and Chairman
Department of Neurobiology
and Anatomy

Stanley G. Schultz, M.D.

Professor and Chairman
Department of Physiology and
Cell Biology

University of Texas Medical School
Houston, Texas

Raven Press New York

Raven Press, Ltd., 1185 Avenue of the Americas, New York, New York 10036

Made in the United States of America

Library of Congress Cataloging-in-Publication Data

Byrne, John H.
    An introduction to membrane transport and bioelectricity :
  (foundations of general physiology and electrochemical signaling) /
  John H. Byrne, Stanley G. Schultz. — 2nd ed.
      p.   cm. — (Raven Press series in physiology)
    Includes bibliographical references and index.
    ISBN 0-7817-0181-3 (hard). — ISBN 0-7817-0201-1 (pbk.)
    1. Electrophysiology.   2. Action potentials (Electrophysiology)
  3. Biological transport.   I. Schultz, Stanley G.   II. Title.
  III. Series.
    [DNLM: 1. Cell Membrane—physiology.   2. Electrophysiology.
  3. Biological Transport.   4. Synaptic Transmission.   QS 532.5.M3
  B9951   1994]
  QP341.B95   1994
  574.87′5—dc20
  DNLM/DLC
  for Library of Congress   94-2556

9 8 7 6 5 4 3 2 1

Contents

Preface to the First Edition

This book is intended to introduce medical and beginning graduate students to the principles of membrane transport and bioelectricity. These principles form the necessary foundation for the understanding of many physiological processes and systems, as well as their regulation. Many presentations of this material rely heavily on mathematical and physiochemical approaches that make these concepts difficult for many students to grasp and comprehend. Consequently, in order to make this material more accessible, we have attempted to present it in such a way that key principles and concepts can be understood with minimal training in the mathematical and physical sciences. Nonetheless, we believe that this presentation sufficiently provides students with the prerequisite background for medical school courses in physiology and the neurosciences, as well as more advanced graduate courses. Formal derivations of all the mathematical expressions in this text may be found in the book entitled *Basic Principles of Membrane Transport*, written by one of us (S.G.S.) and cited at the ends of the appropriate chapters. Throughout this text we maintain a historical perspective, including discussion of some of the experimental strategies. We believe that this perspective helps students understand the material and appreciate the scientific evolution of the field, as well as providing more interesting reading. . . .

This book evolved as a result of our teaching this material to first-year medical students at the University of Pittsburgh School of Medicine and the University of Texas Medical School in Houston during the past two decades. We are grateful to all of our students and colleagues, past and present, for their many helpful comments and suggestions. In particular, we would like to thank Drs. D. Baxter, L. Cleary, and K. Scholz for their comments on the manuscript, and are especially grateful to Dr. Fran Ganong for his detailed, constructive review of this text. Finally, we thank D. Stickle, D. Moss, and J. Pastore for the preparation of the illustrations.

John H. Byrne
Stanley G. Schultz
Houston, Texas

Preface

This Second Edition has been expanded significantly to reflect the many new developments in the areas of membranes and synaptic transmission. Indeed, it is fair to say that during the past five years recent molecular biological approaches and patch clamp techniques have revolutionized the understanding of membrane channels. Thus, we have increased the coverage of this material throughout the book, including the addition of a new chapter on ion diffusion through biological membranes. The coverage of principles of synaptic transmission in the central nervous system and synaptic plasticity has been expanded as well, due to enhanced understanding of these phenomena, as well as the growing realization of their essential role in neuronal signaling and in information processing and storage. This new material, coupled with the earlier material on the mechanisms of action potentials and synaptic transmission, will provide readers with a solid foundation for more advanced studies in these areas.

The First Edition introduced the principles of membrane transport and bioelectricity to medical students and to graduate students. Our objective here is the same, but we have taken the opportunity to eliminate redundancy in some areas, to expand coverage in other areas, and to include additional material for more advanced students. We have eliminated the study questions, added more advanced material on ion permeation, expanded the coverage of osmosis and synaptic transmission, and expanded the Appendix to include a description of some of the specialized membrane conductances and mechanisms that have specific roles in determining the firing pattern and integrative action of individual neurons.

Again we are indebted to our students and colleagues for their thoughtful comments on ways to improve the first edition and the new sections of the second edition. In particular, we would like to thank D. Baxter, L. Cleary, P. Kelly, M. Mauk, and N. Waxham for their comments and T. Vicknair and J. Pastore for assistance with the illustrations.

John H. Byrne
Stanley G. Schultz
Houston, Texas

1

Membrane Composition and Structure

HISTORICAL PERSPECTIVES: THE DAWNING OF MEMBRANE BIOLOGY

The first suggestion of the existence of biological membranes is generally attributed to the botanist Carl Wilhelm Nägeli (1817–1891), who was a pioneer in the application of microscopic techniques to the study of cell detail in an attempt to relate structure and function. In his classic work *Primordialschlauch*, published in 1855, he pointed out that the region immediately adjacent to the inner surfaces of the walls of plant cells appeared to differ from the underlying protoplasm and conjectured that this may be because the protoplasm becomes a firmer gel on contact with the extracellular fluid. Two decades later, another botanist, W. Pfeffer, noted that, when plant cells were placed in concentrated solutions, the protoplasm shrank away from the wall and appeared to be surrounded by a distinct structure, or "membrane." In his treatise dealing with these studies (*Osmotische Untersuchungen*), he argued that these membranes were discrete structures that could serve as selective barriers for the passage of substances into and out of cells. This was merely a guess, but one that had the virtue of being correct; that guess is sometimes referred to as Pfeffer's postulate.

It was not until 1890 that Ernst Overton confirmed Pfeffer's guess. Overton compared the solubility of a large number of solutes in olive oil with the ease with which they permeate (enter) cells and, as is discussed in greater detail in the following sections, concluded that "Pfeffer's membrane" behaved very much as if it were made up of lipids similar to olive oil and thus differed from the aqueous protoplasm.

This first clue to the composition of biological membranes takes us back many years, and, indeed, it would not be too much of an exaggeration to say that the earliest contribution in this area can be attributed to Pliny the Elder (A.D. 23–79) who noted, "Everything is soothed by oil, and this is the reason why divers send out small quantities of it from their mouths, because it smooths every part which is rough" (*Natural History*, Book II, Section 234). Apparently, pearl divers made a habit of diving with a mouthful of oil when the surface of the water was rippled by winds; when they ejected the oil, it would "soothe the troubled waters" and increase underwater visibility.

In 1762, Benjamin Franklin was purportedly reminded of this "calming effect" of oil by an old sea captain, and, in 1765, when he was the American ambassador to the Court of St. James, he performed an experiment that is a landmark in the science of surface chemistry. Franklin poured olive oil onto a pond in Clapham Common (near London) and noted that the ripples caused by the wind were indeed calmed. He also, astutely, noted that a given quantity of oil would only spread over a definite area of the pond; the same quantity of oil would always cover the same area, and twice that volume of oil would cover twice that area. Knowing the volume of oil poured on the pond and the area covered, Franklin calculated that the thickness of the layer formed by the oil was about 25 Å (2.5nm) and that it "could not be spread any thinner." Current measurements of the thickness of a monomolecular layer (Fig. 1.1A) of olive oil on water have not significantly improved on Franklin's estimate.

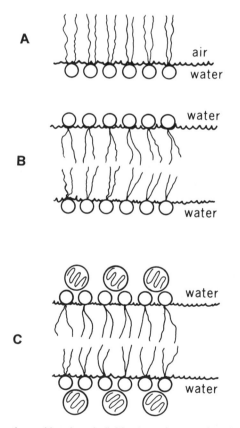

FIG. 1.1. **(A)** Monolayer formed by phospholipids at an air–water interface. **(B)** A phospholipid bilayer separating two aqueous compartments. **(C)** A bimolecular lipoprotein membrane.

In 1925, Gorter and Grendel, stimulated by the findings of Franklin and Overton, performed a series of studies that had a major impact on all subsequent thinking dealing with membrane structure. These investigators extracted the lipids from erythrocyte membranes of a variety of species and calculated the area covered by these lipids when spread on water to form a monomolecular layer. They also approximated the total area of the membranes from which the lipid was extracted and concluded that the area of the monomolecular layer was twice the membrane area; that is, there was sufficient lipid to form a double or bimolecular layer of lipid around the cells, with each layer about 25 Å thick (Fig. 1.1B).

In 1935, Davson and Danielli modified the model proposed by Gorter and Grendel by including protein in the membrane structure. The bimolecular lipoprotein model they proposed is illustrated in Fig. 1.1C. The essential features of this model are (1) a bimolecular lipid core that is 50 Å (5nm) thick and corresponds to the Gorter-Grendel model; and (2) inner and outer protein layers attached to the polar head groups of the lipids by ionic interactions. This model enjoyed 25 years of essentially universal acceptance.

CURRENT CONCEPTS: THE "FLUID-MOSAIC" MEMBRANE

Since 1945, the application of increasingly sophisticated analytical and ultrastructural techniques to the study of the composition and structure of biological membranes has met with remarkable success.

First, the early notion that biological membranes are made up of a mixture of lipids and proteins has been firmly established for all such barriers throughout the animal and plant kingdoms. The proportions of these two components differ among different cell membranes. In general, membranes that primarily serve as insulators between the intracellular and extracellular compartments and have few metabolic functions (e.g., myelin) are made up of a relatively high proportion of lipids compared to proteins. On the other hand, membranes that surround "metabolic factories" (e.g., hepatocytes, mitochondria) are relatively rich in protein content compared to lipid content.

Second, it is now generally accepted that the lipids that comprise biological membranes are primarily from the group referred to as phospholipids. While cholesterol is present in the membranes of many eurokaryotic cells, it is not found in most prokaryotic cells. Now, all phospholipids are derivatives of phosphatidic acid, which consists of a phosphorylated glycerol "back-bone" to which two fatty acid "tails" are attached by ester bonds. The most prevalent phospholipids found in biological membranes, such as phosphatidylcholine, phosphatidylserine, phosphatidylinositol, and phosphatidylethanolamine, result from the esterification of the free phosphate group of phosphatidic acid with the hydroxyl groups of choline, serine, inositol, and ethanolamine, respectively. The important point is that all phospholipids contain a water-soluble (hydrophilic) "head-group" (i.e., the phosphorylated glycerol back-bone and its conjugates) and two water-insoluble (hydrophobic or lipophilic) "tails." Such molecules are referred to as amphiphatic (or "amphipathic"). [The prefix "amphi" derives from both the Greek and Latin, meaning "having two sides," e.g., *amphibians* live in air (on land) and in water]. When these compounds are poured onto water, their hydrophilic head groups will enter the aqueous phase and their hydrophobic tails will simply wave in the air (Benjamin Franklin's observation; see Fig. 1.1A). But, when these molecules are confronted with two aqueous compartments, they will spontaneously form a bilayer (i.e., the Gorter and Grendel model; see Fig. 1.1B) with their water-soluble head-groups immersed in the two aqueous phases and their lipid tails, forming a hydrophobic core.

Finally, perhaps the most revolutionary advance in this area has been in our understanding of the way in which these proteins and lipids are assembled in biological membranes. It is now clear that the phospholipids and cholesterol form an "oily" fluid bilayer in which the adherent proteins are free to float around at will. Some of these proteins span the thickness of the bilayer; these so-called integral proteins have hydrophobic middles and hydrophilic ends, so that their ends protrude into the intracellular and extracellular watery compartments, while their middles are "glued" by hydrophobic bonds within the oil (Fig. 1.2). Other proteins are electrostatically attached to either the inner or outer surfaces of the bilayer; they are referred to as "peripheral proteins." In stark contrast with the Davson-Danielli "biomolecular lipoprotein" model (Fig. 1.1C), which has a rather static appearance, this fluid-mosaic model is dynamic and possesses structural features that afford avenues for communication between the extracellular milieu and the intracellular compartments by virtue of signal transduction and the selective exchange of solutes. In short, the fluid-mosaic model provides an ideal basis for the correlation of membrane structure and function.

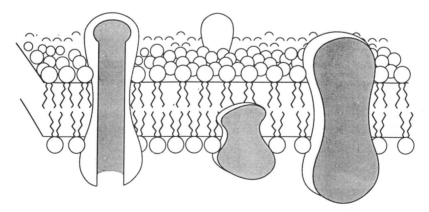

FIG. 1.2. Schematic representation of the fluid-mosaic model proposed by Singer and Nicholson, featuring a phospholipid bilayer containing peripheral and integral proteins.

BIBLIOGRAPHY

Bretscher MS. The molecules of the cell membrane. *Sci Am* 1985;253(4):100–108.

Darnell J, Lodish H, Baltimore D. *Molecular cell biology*. Second Edition New York: Scientific American Books, Inc., 1990; Chapter 13.

Singer SJ, Nicholson GL. The fluid mosaic model of the structure of cell membranes. *Science* 1972; 175:720–731.

2

Diffusion of Nonelectrolytes

The term *diffusion* refers to the net displacement (transport) of matter from one region to another due to random thermal motion. It is classically illustrated by an experiment in which an iodine solution is placed at the bottom of a cylinder and pure water is carefully layered above this colored solution. Initially, there is a sharp demarcation between the two solutions; however, as time progresses, the upper solution becomes increasingly colored, and the lower solution becomes progressively pale. Ultimately, the column of fluid achieves a uniform color, and the diffusion of iodine ceases. This state of maximum homogeneity, or uniformity, will persist as long as the cylinder is undisturbed; a perceptible deviation from homogeneity on the part of the entire column or any bulk (macroscopic) portion of the column will never be observed. The properties of the system, at this point, are said to be time-independent; or, stated in another way, the system is said to have achieved a state of *equilibrium*.

The kinetic characteristics of diffusion can be readily developed by considering the simplified case illustrated in Fig. 2.1. Compartment o and compartment i both contain aqueous solutions of some uncharged[1] solute (i) at concentrations C_i^o and C_i^i, respectively, where the superscripts designate the compartment and the subscript designates the solute.

These compartments are separated by a sintered glass disk, which, because it possesses many large pores, may be treated as if it were an aqueous layer having a cross-sectional area A and a thickness Δx; the disk simply serves as a barrier to prevent bulk mixing of the two solutions. We assume that each compartment is well stirred so that the concentrations of C_i^o and C_i^i are uniform. Furthermore, for the sake of simplicity, we will also assume that both compartments have sufficiently large volumes so that their concentrations remain essentially constant, during the period of observation, despite the fact that the solute i may leave or enter these compartments across the disk.

Because the solute molecules are in continual random motion, due to the thermal

---

[1]An uncharged solute refers to a molecule that bears no net charge and thus includes solutes that have an equal number of oppositely charged groups (e.g., zwitterions).

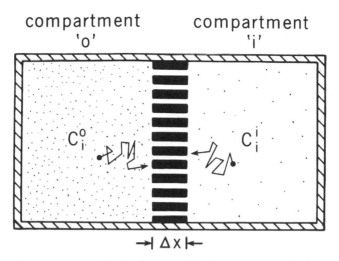

FIG. 2.1. Random movements of solute *i* across a highly porous sintered disk having a thickness of Δx.

energy of the system, there is a continual migration of these molecules across the disk in both directions. Thus, some of the molecules originally in compartment o will randomly wander across the barrier and enter compartment i. The rate of this process, because it is the result of random motions, is proportional to the likelihood that a molecule in compartment o will enter the opening of a pore in the disk and is therefore proportional to C_i^o. Thus we may write

Rate of molecular migration (diffusion) from o to i $= kC_i^o$

where k is a proportionality constant.

Similarly, we may write

Rate of molecular migration (diffusion) from i to o $= kC_i^i$

The rate of net movement of molecules across the barrier is the difference between the rates of these two unidirectional movements so that

Rate of net molecular migration (diffusion) across barrier $= k(C_i^o - C_i^i) = k\Delta C_i$

Thus, the net flow of an uncharged solute across a permeable barrier due to diffusion is directly proportional to the concentration difference across the barrier. In addition, the rate of transfer is directly proportional to the cross-sectional area A and inversely proportional to the thickness of the barrier Δx; that is, for a given concentration difference ΔC_i, doubling the area of the disk will result in a doubling of the rate of transfer from one compartment to the other, whereas doubling the thickness of the disk will halve the rate of transfer. Therefore, we may replace k with a more explicit expression that contains a new proportionality constant D_i, where $k = AD_i/\Delta x$, and obtain

Rate of net migration (diffusion) across the barrier $= AD_i \Delta C_i / \Delta x$

Dividing both sides of this expression by A, we obtain the rate of net flow per unit area, which is often referred to as the flux (or net flux). This is commonly symbolized by J_i, which is expressed in units of amount of substance, per unit area, per unit time (e.g., mol/cm^2 hr). Thus,

$$J_i = D_i \Delta C_i / \Delta x \qquad [2.1]$$

where D_i is the diffusion coefficient of the solute i and is a measure of the rate at which i can move across a barrier having an area of 1 cm^2 and a thickness of 1 cm when the concentration difference across this barrier is 1 mol/liter. The coefficient D_i is dependent on the nature of the diffusing solute and the nature of the barrier or the medium in which it is moving (interacting); we shall examine this dependence in further detail below. If J_i is expressed in mol/cm^2 hr, Δx in cm, and ΔC_i in mol/1,000 cm^3, then D_i emerges with the (somewhat uninformative) units of cm^2/hr.

Equation 2.1 describes diffusion across a flat or planar barrier having a finite thickness. Such systems are often called discontinuous systems, inasmuch as the barrier introduces a sharp and well-defined demarcation between the two surrounding solutions. Diffusion of iodine into water, in the experiment described above, represents a continuous system as there is no discrete boundary between the two solutions. The general expression for diffusion in a continuous system can be derived from Eq. 2.1 by simply making the thickness of the disk vanish mathematically. Thus, as Δx approaches 0, $\Delta C_i / \Delta x$ approaches dC_i/dx so that

$$J_i = D_i [dC_i/dx] \qquad [2.2]$$

Equation 2.2 was derived in 1855, by Fick, a physician, and is often referred to as *Fick's (first) law of diffusion*. It simply states that the rate of flow of an uncharged solute due to diffusion is directly proportional to the rate of change of concentration with distance in the direction of flow. The derivative dC_i/dx is referred to as the *concentration gradient* and is the "driving force" for the diffusion of uncharged particles.

Thus, Eq. 2.2 states that there is a linear relation between the diffusional flow of i and its driving force, where D_i is the proportionality constant. Equation 2.2 is but one example of many linear relations between flows and driving forces observed in physical systems. For example, Ohm's law states that there is a linear relation between electrical current (flow) and its driving force, the voltage (electrical potential difference), where the proportionality constant is the electrical conductance g (recall that $g = 1/R$, where R is resistance). Thus, we can write

$$I = gV \text{ or } IR = V \qquad [2.3]$$

We will see below that the diffusion of charged particles (ions) can be described by a combination of Eqs. 2.2 and 2.3, where the driving force is a combination of the concentration difference (chemical force or potential) and the electrical potential difference.

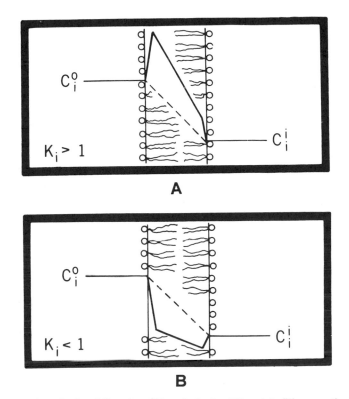

FIG. 2.2. Partitioning of a lipophilic solute **(A)** and a hydrophilic solute **(B)** across the interfaces of a lipid membrane. The solid lines illustrate concentration profiles in the two solutions and within the lipid membrane.

DIFFUSION THROUGH SELECTIVE BARRIERS

A sintered glass disk is highly porous, and the dimensions of the water-filled pores that penetrate the disk are extremely large compared with molecular dimensions. For this reason, the diffusing solute molecules behave as if they were passing through an aqueous layer and are unaffected by the presence of the barrier; in other words, they are not significantly influenced by either the openings or the walls of the pores through which they pass. The diffusion coefficient of a solute within the disk is the same as its diffusion coefficient in a continuous aqueous system ("free solution"), and the concentration of the solute just within the disk at each interface is the same as its concentration in the immediately adjacent aqueous solution. Thus, the sintered glass disk is a nonselective barrier because the properties of the solute within the disk are the same as those in the surrounding solutions.

If, as illustrated in Fig. 2.2, the sintered glass disk is replaced by a lipid membrane, the description of diffusion becomes slightly more complicated.

First the concentration of the solute just within the membrane at each interface, in general, will not be equal to the concentration in the immediately adjacent aqueous solutions. If the solute is hydrophilic (or lipophobic; e.g., an ion), it will prefer the aqueous phase to the lipid phase, and its concentration just within the membrane will be less than that in the adjacent aqueous solutions. On the other hand, if the solute is lipophilic (e.g., fats, sterols, phospholipids), it will distribute itself so that the concentration at the interfaces just within the lipid barrier will exceed that in the adjacent aqueous solutions. The distribution of a given solute between the aqueous phase and the adjacent oil or lipid phase is described, quantitatively, by a unitless number termed a *partition coefficient* (also, a *distribution coefficient*). The partition coefficient of a given solute i between two solvent phases is determined in the following manner: The two immiscible solvents, for example, olive oil and water, together with an arbitrary amount of solute are placed in a separatory funnel. The funnel is then stoppered and shaken until an equilibrium distribution of the solute i in the two phases is achieved. The olive oil–water partition coefficient is defined as

$$K_i = \frac{\text{Equilibrium concentration of } i \text{ in olive oil}}{\text{Equilibrium concentration of } i \text{ in water}}$$

In this simple way, partition coefficients between lipid and aqueous phases have been determined for a large variety of solutes.

Returning to the example illustrated in Fig. 2.2A, it follows that if the solute in the two compartments is hydrophobic or lipophilic ($K_i > 1$), the concentration just within the membrane at the interface with compartment o will be $K_i C_i^o$, and the concentration just within the membrane at the interface with compartment i will be $K_i C_i^i$. Thus, the concentration difference within the membrane is $K_i \Delta C_i$ so that the partition coefficient has essentially amplified the driving force for the diffusion of the solute across the membrane. The flux of solute through the lipid membrane due to diffusion is now given by

$$J_i = K_i D_i \Delta C_i / \Delta x \qquad [2.4]$$

where the diffusion coefficient D_i now reflects the ease with which the solute can move through the lipid and will differ from the value of D_i in an aqueous solution.[2]

Conversely, if the solute is highly water soluble (i.e., hydrophilic or lipophobic), then $K_i < 1$ and its concentration difference within the membrane will be attenuated.

Now, in general, K_i, D_i, and Δx cannot be readily determined. For this reason, these unknowns have been lumped together to give a new term, the *permeability coefficient*, P_i, which is defined as

$$P_i = K_i D_i / \Delta x \qquad [2.5]$$

Since K_i is unitless, when D_i is given in cm^2/hr and Δx in cm, then P_i has units of cm/hr. The permeability coefficient of a given membrane for a given solute is simply

---

[2] We are assuming that diffusion through the membrane is slow (i.e., rate-limiting) compared to partitioning into and out of the membrane at the two interfaces.

$$P_i = J_i/\Delta C_i \qquad\qquad [2.6]$$

and can be readily determined experimentally. Clearly, the permeability coefficient of a membrane for an uncharged solute is the flow of solute (in moles per hour) that would take place across 1 cm^2 of membrane when the concentration difference across the membrane is 1 M.

We now recapitulate some of the laws that apply to diffusional movements across artificial as well as biological membranes.

First and foremost, the driving force or *sine qua non* for the net diffusion of an uncharged solute is a concentration difference. In the absence of a concentration difference, a net flux due to diffusion is impossible. In the presence of a concentration difference, diffusion will take place spontaneously and the direction of the net flux is such as to abolish the concentration difference. In other words, diffusion only brings about the transfer of net uncharged solutes from a region of higher concentration to a region of lower concentration; the reverse direction is thermodynamically impossible! For this reason, transport due to diffusion is often referred to as "downhill" (because the flow is from a region of higher concentration to one of lower concentration) or "passive" transport (because no additional energy need be supplied to a system to enable these flows to take place; the inherent thermal energy responsible for random molecular motion is sufficient). As we shall see, there are numerous biological transport processes that bring about the flow of uncharged solutes from a region of lower concentration to one of higher concentration. These "uphill," or "active," transport processes cannot be due to diffusion and are dependent on an energy supply in addition to simple thermal energy.

Within the membrane, the link between the driving force ΔC_i and the flow J_i is the diffusion coefficient D_i. This is the factor that determines the flow per unit driving force, and it is determined by the properties of the diffusing solute and those of the membrane through which diffusion takes place. At the molecular level, the diffusion coefficient is a measure of the resistance offered by the membrane to the movement of the solute molecule.

At the turn of the century, Einstein (1905) proposed that the resistance experienced by a diffusing particle results from the frictional interaction between the surface of the particle and the surrounding medium because the two, in essence, are moving relative to each other. Drawing on Stoke's law, which describes the friction experienced by a sphere falling through a medium having a given viscosity, Einstein demonstrated that the diffusion coefficient of a spherical molecule should be inversely proportional to both the radius of the molecule and the viscosity of the surrounding medium. Thus, when one is concerned with the diffusion of a variety of solutes through a single medium (fixed viscosity), the relative diffusion coefficients should be inversely proportional to their molecular radii. Einstein's contributions to our understanding of diffusion and the nature of the diffusion coefficient have been repeatedly confirmed, and the Stokes-Einstein equation has proved to be a valuable approach to the calculation of molecular dimensions from measurements of diffusion coefficients.

We can now make several educated guesses about the permeability of biological membranes to uncharged solutes: First, because biological membranes are primarily composed of lipids, Eq. 2.5 suggests that the permeability coefficients of solutes having approximately the same molecular dimensions (same diffusion coefficients) should vary directly with their partition coefficients (K's); that is, the more lipid soluble the molecule, the greater its permeability coefficient. This deduction has been verified repeatedly and is a statement of *Overton's law*. In fact, as already noted, Overton's observation that lipid-soluble molecules penetrate biological membranes more readily than water-soluble molecules of the same size predated the chemical analyses of membranes and was the first clue that biological membranes are made up of lipids.

Second, Eq. 2.5 also predicts that the permeability coefficients of solutes having the same lipid solubilities (same K's) should vary inversely with their molecular sizes; that is, the larger the molecule, the lower its permeability coefficient. It should be noted that K's can vary over many orders of magnitude whereas D's generally do not differ by more than a factor of five. Thus, the permeability coefficients for simple diffusion of uncharged solutes across biological membranes are more strongly influenced by differences in lipid solubility than molecular size.

MEMBRANE PORES AND RESTRICTED DIFFUSION

By the early 1930s, the ability of Eq. 2.4 to describe diffusion across biological membranes was experimentally well established, and the prevalent view was that cells are surrounded by an intact or continuous lipid envelope. This view was soon challenged by the exhaustive studies of Collander and his associates on the permeability of the plant cell *Chara* to a large number of solutes. The results of these studies are summarized in Fig. 2.3, where the permeability coefficients of a number of solutes (ordinate) are plotted against their oil-water partition coefficients (abscissa).

Clearly, there is, in general, a direct linear relationship between these parameters as previously demonstrated by Overton; but Collander also noted that small water-soluble molecules tended to lie above the line describing the best fit of the data. In other words, small water-soluble molecules appeared to permeate the membrane more rapidly than could be accounted for on the basis of their lipid solubilities alone.

This astute observation led Collander to conclude:

> "It seems therefore natural to conclude that plasma membranes of the *Chara* cells contain lipoids, the solvent power of which is on the whole similar to olive oil. But, while medium sized and large molecules penetrate the plasma membrane only when dissolved in the lipoids, the smallest molecules can also penetrate in some other way. Thus, the plasma membrane acts both as a *selective solvent* and as a *molecular sieve*."

In short, he proposed that small, highly water-soluble molecules can penetrate biological membranes through two parallel pathways: (a) They can dissolve in the lipid matrix and diffuse through the membrane just as lipid-soluble molecules do but to a

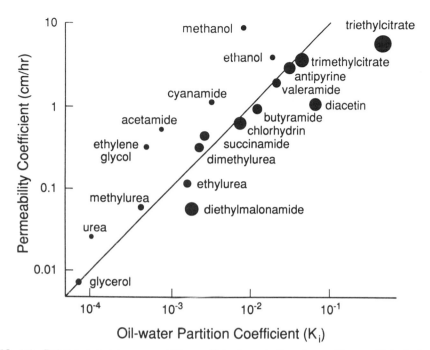

FIG. 2.3. Relation between permeability coefficients and oil–water partition coefficients for a number of solutes. The sizes of the circles are proportional to the molecular diameters of the solutes.

very much lesser extent; and (b) they can pass through channels or pores that perforate the lipid envelope.

During the years that have elapsed since Collander's suggestion of a "mosaic lipid" membrane (i.e., a discontinuous lipid layer perforated by water-filled channels), a compelling body of evidence has accrued for the presence of proteinaceous channels that span the lipid bilayer and serve as the routes for the diffusional flows of highly water-soluble solutes across biological membranes. These flows are often referred to as *restricted diffusion* inasmuch as their rates are dramatically influenced by molecular size.

SUMMARY

We may now summarize the prevalent view of diffusion of uncharged solutes through biological membranes.

1. Lipid-soluble molecules readily penetrate the membrane by diffusion through the lipid bilayer. The permeability coefficients of large lipid-soluble molecules may be quite high due to partition coefficients that strongly favor their entry into the lipid phase. These movements are often referred to as *simple diffusion*.

2. Small water-soluble molecules, which have little affinity for the lipid phase, may penetrate the membrane through pores. This restricted diffusion is, however, limited to small molecules; most cell membranes are essentially impermeable to water-soluble molecules having five or more carbon atoms. Most, if not all, essential metabolic substrates (e.g., glucose, essential amino acids, water-soluble vitamins) cannot enter the cell to any appreciable extent by this mechanism. As we shall see, other mechanisms ("carrier-mediated" processes) are present that mediate the entry of these substances into cells at rates that are sufficiently rapid to sustain essential metabolic processes.

3. Inasmuch as water and small water-soluble uncharged molecules have finite lipid solubilities, it is often difficult to determine experimentally the extent to which they traverse biological membranes by diffusing through the lipid phase and/or by restricted diffusion through pores. It is generally accepted that the permeation rates of water and other small, polar nonelectrolytes (e.g., urea, glycerol) through biological membranes (where determined) are too fast to be accounted for by diffusion through the lipid phase and that pores must be invoked. There is no doubt that inorganic ions traverse biological membranes virtually exclusively via pores inasmuch as their solubilities in lipid are minute.

BIBLIOGRAPHY

Diamond JM, Wright EM. Molecular forces governing nonelectrolyte permeation through cell membranes. *Proc R Soc Lond (Biol)*, 1969;172:273.

Finkelstein A. *Water movement through lipid bilayers, pores and plasma membranes: Theory and reality*. New York; Wiley-Interscience, 1987: Chapters 6 and 10.

Stein WD. *Transport and diffusion across cell membranes*. Orlando: Academic Press, 1986; Chapter 2.

3

Osmosis

Osmosis refers to the flow or displacement of volume across a barrier due to the movements of matter in response to concentration differences. Although, in principle, any substance (solutes as well as solvents) may contribute to the volume of matter displaced during osmotic flow, the term osmosis has come to have a much more restricted meaning when applied to biological systems. Because biological fluids are relatively dilute aqueous solutions in which water comprises more than 95% of the volume, osmotic flow across biological membranes has come to imply the displacement of volume resulting from the flow of water from a region of higher water concentration (a dilute solution) to a region of lower water concentration (a more concentrated solution).

Osmosis and the diffusion of uncharged solutes are closely related phenomena. Both are spontaneous processes that involve the flow of matter from a region of higher concentration to one of lower concentration; both are the results of random molecular movements and, hence, are dependent only on the thermal forces inherent in any system; and the end result of both processes is the abolition of concentration differences.

How do diffusion and osmosis differ? As we shall see, the answer to this question carries with it the key to understanding osmosis.

Let us start by reexamining the definitions of these two processes. *Osmosis* refers to the flow of matter that results in a displacement of volume, and *diffusion* refers to a flow of matter in which displacements of volume are not involved. As it turns out, the key to understanding osmosis is the answer to the question, Why is there no displacement of volume in diffusion?

The answer to this question emerges when we carefully reconsider the molecular events involved in the diffusion of uncharged solutes described in Chapter 2. Referring once more to Fig. 2.1, when we say that the concentration of an uncharged solute in compartment o is greater than that in compartment i, we are, at the same time, implying that the concentration of water (or solvent) in compartment i is greater than that in compartment o. When there is a concentration difference across a membrane for one component of a binary solution (one solute and one solvent), there must also be a concentration difference for the other component. Thus, there

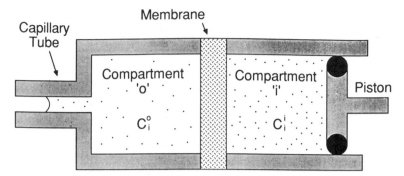

FIG. 3.1. Apparatus for determining osmotic pressure and flow.

are two concentration differences, two driving forces, and two diffusional flows; solute diffuses from compartment o to compartment i, and water diffuses from compartment i to compartment o. In short, when we say that there is diffusion of a solute down a concentration difference we are describing only one-half of the mixing process and are overlooking the fact that diffusion (or mixing) in the closed system illustrated in Fig. 2.1 is, in fact, interdiffusion. Clearly, in the closed system illustrated in Fig. 2.1, if the membrane is rigid, mixing or interdiffusion of solute and solvent must take place without any change in the volumes of compartments o or i.

To appreciate how diffusion can result in the displacement of volume, let us consider the system illustrated in Fig. 3.1. Compartment o is open to the atmosphere and contains pure water. Compartment i is closed by a movable piston and contains an aqueous solution of some uncharged solute. The membrane separating the two compartments is assumed to be freely permeable to water but impermeable to the solute (i.e., a semipermeable membrane). Although there are two concentration differences, there can be only one flow. Water will flow from compartment o to compartment i, driven by its concentration difference. The volume of water associated with this flow, however, cannot be counterbalanced by a flow of solute, so that there will be a net displacement of volume; the volume of compartment o will decrease, whereas that of compartment i will increase. This volume flow, referred to as osmosis or osmotic flow, arises because the properties of the barrier or membrane are such as to prevent interdiffusion; mixing, which is the end to which all spontaneous processes are directed, can now only come about as the result of one flow rather than two flows.

Now, we can prevent the flow of volume from compartment o to compartment i by applying a sufficient pressure on the piston. The pressure that must be applied to prevent the flow of volume is defined as the *osmotic pressure*. When the solutions on both sides of the membrane are relatively dilute (as in the case of biological fluids) and when the membrane is absolutely impermeable to the solute (i.e., interdiffusion is completely prevented), the osmotic pressure is given by the following expression, which was derived by the Dutch Nobel Laureate Jacobus Henricus van't Hoff (1852–1911):

$$\Delta\pi = RT\,\Delta C_i \qquad\qquad [3.1]$$

where $\Delta\pi$ denotes the osmotic pressure; R is the gas constant; T is the absolute temperature; and ΔC_i is the concentration difference of the impermeant, uncharged solute across the membrane in moles per liter. At 37°C ($T = 310°K$), this expression reduces to

$$\Delta\pi = 25.4\,\Delta C_i \text{ (atm)}$$

Thus, if the concentration difference across the membrane is 1 mol/liter, the pressure that must be applied on the piston to prevent osmotic flow of water is 25.4 atm.

At this point, we digress briefly to consider the full meaning of the term ΔC_i in Eq. 3.1. The ability of a solution of any solute to exert an osmotic pressure across a semipermeable membrane [i.e., a membrane permeable to the solvent (water) but ideally impermeable to the solute] is a *colligative property* of the solution that is dependent on the concentration of individual solute particles (other such colligative properties are the vapor-pressure depression, boiling-point elevation, and freezing-point depression of solutions). We must therefore introduce a new unit of concentration that is a measure of the number of free particles in the solution and thus reflects the osmotic effectiveness of a dissolved solute. For a nondissociable solute such as urea, whose molecular weight is 60, when we dissolve 60 g (i.e., 1 gram weight) of this solute in a sufficient amount of water to yield a total volume of 1 liter, we have a 1-M solution of urea that contains about 6×10^{23} individual particles (Avogadro's number). However, when we dissolve 58 g NaCl (whose molecular weight is 58) in a sufficient amount of water to yield a total volume of 1 liter, we obtain a 1-M solution of NaCl; but the number of individual particles in solution will be twice that of a 1-M solution of urea. Thus, one must distinguish between the molarity (the number of gram weights of a solute in a liter of aqueous solution) and the osmolarity (the number of individual particles per liter that results from dissolving one gram weight of a solute in that volume). A 0.15-M solution of NaCl that consists of 0.15 mol/liter Na^+ particles and 0.15 mol/liter Cl^- particles has the same osmolarity as a 0.3-M sucrose solution (i.e., 300 mosm/liter). Indeed, historically, this observation provided the crucial evidence for the Arrhenius theory of the dissociation of salts in solution into their constituent ions.

In short, the osmolarity or the osmotically effective concentration of a solution of a dissociable salt will be n times its molarity, where n represents the number of individual ions (particles) resulting from the dissociation of the salt.

Finally, before considering osmotic flow across biological membranes, it is important to gain deeper insight into the nature of osmotic pressure. When, as in the example given above, sufficient pressure is applied to the piston to prevent volume flow, the system is in equilibrium. There will be no flow of water despite the presence of a concentration difference for water across the membrane. This is because the pressure applied to compartment i exerts the same driving force for the flow of water as does the difference in water concentration, but in the opposite direction. Therefore, when the osmotic pressure is applied, there is no longer a net driving force for the movement of water, and volume flow ceases.

One may find this easier to understand by arbitrarily dividing the water flows into two hypothetical streams. One stream is directed from compartment o into compartment i and derives its driving force from the difference in the concentration of water across the membrane. The second stream is directed from compartment i to compartment o and is driven by the pressure applied by the piston. Flow ceases when these two oppositely directed streams are of equal magnitude; the pressure needed to achieve this equilibrium state is given by van't Hoff's law (Eq. 4.1). This balancing of driving forces is analogous to the condition discussed in Chapter 4 with reference to the Nernst equilibrium potential, where an electrical potential difference counterbalances the driving force arising from a concentration difference of a permeant ion when the counterion is impermeant.

OSMOTIC FLOW AND OSMOTIC PRESSURE ACROSS NONIDEAL MEMBRANES

Up to this point, we have considered the case of osmotic flow across a semipermeable membrane, that is, one that is permeable to solvent but impermeable to the solute so that interdiffusion is completely prevented. What would happen if the membrane in Fig. 3.1 could not distinguish at all between a solvent molecule and a solute molecule, that is, both could cross with equal ease? Suppose, for example, that compartment o contains pure water so that the concentration of water in that compartment $C_{H_2O}^o = 55.6$ M. And, suppose that compartment i contains a 1-M solution of deuterium oxide (D_2O or "heavy water") in water. Under these circumstances $C_{H_2O}^i \approx 54.6$ M so that $\Delta C_{H_2O} \cong \Delta C_{D_2O} = 1$ M. Thus, H_2O would diffuse from compartment o to compartment i driven by a concentration difference of 1 M, and D_2O would diffuse from compartment i to compartment o driven by an equal but oppositely directed concentration difference. Since both species can cross the membrane with equal case, interdiffusion or mixing would not be restricted and would take place with no displacement of volume. In short, when the membrane cannot distinguish between solute and solvent, osmotic flow, $J_v = 0$ and, obviously, $\Delta \pi = 0$.

Clearly, the situation in which the membrane is ideally impermeable to the solute and the one in which the membrane cannot distinguish between solute and solvent are extreme examples. In most instances, we are faced with a situation somewhere between these two extremes, that is, a situation in which interdiffusion is to some extent, but not completely, restricted. Under this condition, for a given concentration difference the volume displacement, or osmotic flow, will be somewhere between zero and the maximum that would be observed with an ideal semipermeable membrane. Also, the osmotic pressure (the pressure necessary to abolish osmotic flow) will be between zero and that predicted by van't Hoff's law.

In 1951, the Dutch physical chemist Staverman provided a quantitative expres-

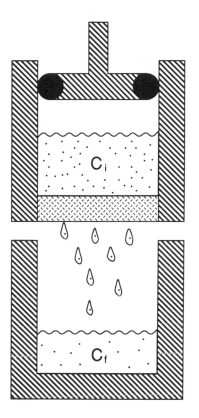

FIG. 3.2. Determination of the reflection coefficient of a solute *i* employing ultrafiltration through a membrane.

sion for the osmotic pressure across nonideal membranes, which was based on the following reasoning: Let us perform the ultrafiltration experiment illustrated in Fig. 3.2. In the upper cylinder, we place an aqueous solution of a solute at a concentration of C_i. We then apply pressure to the piston, force fluid through the membrane, and collect the filtrate. If the membrane does not restrict the movement of solute relative to that of water, then the concentration of solute in the filtrate, C_f, will be equal to C_i; that is, both components passed through the membrane in the same proportion as they existed in the solution of origin. At the other extreme, if the membrane is impermeable to the solute, the filtrate will be pure water (i.e., $C_f = 0$). Between these two extremes is the condition where the membrane partially restricts the movement of the solute relative to that of water, and under these conditions C_f will be lower than C_i but greater than zero. We can now define a parameter that tells us something about the relative ease with which water and the solute *i* can traverse the membrane:

$$\sigma_i = 1 - \{C_f/C_i\}$$

where σ_i is the *reflection coefficient* of the membrane to *i*, because it is a measure of the ability of the membrane to "reflect" the solute molecule *i*; that is, it tells us how perfect the membrane is as a "molecular sieve." Clearly, the reflection coefficient must have a value between zero (for the case where the membrane does not distinguish between the solute and water) and unity (for a membrane that is absolutely impermeant to the solute). Staverman then showed that if the same membrane used in the ultrafiltration experiment is mounted in the apparatus shown in Fig. 3.1, and if two solutions of the same solute at two different concentrations are placed in compartments o and i, the *effective osmotic pressure* necessary to prevent volume flow is given by

$$\Delta \pi_{eff} = \sigma_i \, RT \, \Delta C_i \qquad\qquad [3.2]$$

Because σ_i must have a value between zero and unity, the effective osmotic pressure across a real membrane must fall between zero and that predicted by van't Hoff's law for ideal semipermeable membranes; the exact value depends on the concentration difference and the relative permeability of the membrane to water and the solute i given by σ_i.

In short, Eq. 3.2 permits quantitation of the effect of interdiffusion between solutes and solvent (water) across membranes on the effective osmotic pressure that is exerted across those membranes. If a membrane is equally permeable to the solvent (water) and the solute (i), then $\sigma_i = 0$, and the presence of a concentration difference for i across the membrane will not generate an osmotic pressure. Conversely, if the membrane is impermeant to i, then interdiffusion is prohibited, and the effective osmotic pressure across the membrane will be given by van't Hoff's equation.

VOLUME FLOW IN RESPONSE TO A DIFFERENCE IN PRESSURE

Volume flow, J_v, across a membrane in response to a difference in hydrostatic pressure, ΔP, is given by the linear relation

$$J_v = K_f \, \Delta P \qquad\qquad [3.3]$$

where K_f is a proportionality constant referred to as the hydraulic conductivity of the membrane (or, the filtration coefficient); Eq. 3.3 is analogous to Fick's law of diffusion (Eq. 2.2) inasmuch as it describes a linear relation between a flow and its driving force, which in the case of volume flow is the pressure difference across the barrier. Now, ΔP can be the difference in hydrostatic pressure, the difference in osmotic pressure, or a combination of both. For example, referring to Fig. 3.1, if the concentration in compartment o is equal to that in compartment i so that $\Delta \pi = 0$, application of pressure to the piston will bring about the flow of volume from i to o given by Eq. 3.3. Alternatively, if $C_i^o < C_i^i$, and no pressure is applied to the piston, there will be a flow of volume from o to i given by

$$J_v = K_f \sigma_i RT(C_i^o - C_i^i) \qquad\qquad [3.4]$$

An important empirical observation made many years ago is the equivalence of osmotic and hydrostatic pressure as the driving forces for volume flow. That is, J_v

across a given membrane will be the same when it is driven by a hydrostatic pressure difference as when driven by the equivalent osmotic pressure difference. In other words, for a given membrane, the same value of K_f applies to both forces. Thus, we can combine Eqs. 3.3 and 3.4 and derive a general equation that describes the situation when there are both osmotic and hydrostatic pressure differences across the membrane:

$$J_v = K_f(\Delta\pi_{eff} - \Delta P) \qquad [3.5]$$

where $\Delta\pi_{eff} = \sigma_i RT(C_i^o - C_i^i)$.

Thus, when $\Delta P = \Delta\pi_{eff}$, $J_v = 0$; this is the definition of osmotic pressure. When $\Delta P \neq \Delta\pi_{eff}$, there will be a flow from one compartment to the other, driven by the difference.

Equation 3.5 provides a general description of the effects of hydrostatic and osmotic forces on volume flow across all membranes. In the physiological sciences, it is often referred to as the Starling equation, after the great British physiologist Ernst Starling, who applied it to the study of fluid movements across the walls of capillaries.

PATHWAYS FOR WATER FLOW ACROSS BIOLOGICAL MEMBRANES

As discussed in Chapter 2, solute molecules can cross biological membranes by partitioning (or dissolving) into and diffusing through the lipid bilayer and/or by diffusing through aqueous pathways that traverse the barrier. The same is true for water; indeed, lipid bilayers may be more permeable to water than to more polar nonelectrolytes such as urea. Now, all of the relations described by Eqs. 3.1–3.4 are based on thermodynamic or "quasi-thermodynamic" reasoning and are true (or almost true) regardless of the pathways for solute and water flow. But the physical connotations of K_f and σ_i do depend upon the pathways for these flows. Let us briefly consider this matter qualitatively with the aid of Fig. 3.3. Assume that in all three instances compartment o initially contains pure water and compartment i contains a solution of solute i.

In Fig. 3.3A, the membrane is a pure lipid bilayer and does not possess continuous aqueous pathways for water or solute flow. In this case, water may diffuse from compartment o to compartment i and solute may diffuse in the opposite direction in accordance with Eq. 2.4 or Eq. 2.6. The rate of volume displacement resulting from these flows will simply be the difference between the rates of the volumes of water and solute crossing the barrier in opposite directions.

Now, most biological membranes are much more permeable to water than are simple lipid bilayers due to the presence of water-filled channels or pores. Figure 3.3B depicts a situation in which, in addition to diffusing through the lipid phase of the membrane, both water and solute may also cross the membrane through distinct, selective channels. Once again, the rate of displacement of volume, J_v, is the difference between the rates of the opposite directed volume flows of water and solute.

Compartment "o"   Compartment "i"

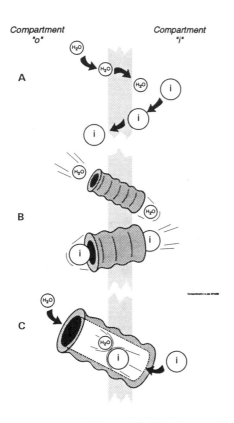

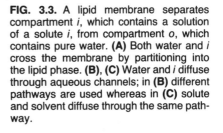

FIG. 3.3. A lipid membrane separates compartment *i*, which contains a solution of a solute *i*, from compartment *o*, which contains pure water. **(A)** Both water and *i* cross the membrane by partitioning into the lipid phase. **(B)**, **(C)** Water and *i* diffuse through aqueous channels; in **(B)** different pathways are used whereas in **(C)** solute and solvent diffuse through the same pathway.

In cases 3.3A and 3.3B, the oppositely directed flows of water and solute take place through parallel pathways and do not interact. If, however, solute and water diffuse across the membrane in opposite directions through *common* pathways, as depicted in Fig. 3.3C, the situation may be somewhat more complicated. In this instance, the rate of volume displacement from compartment o to compartment i will be affected by the frictional interactions between these two oppositely directed flows. This "rubbing-up" between water and solute molecules will slow their respective flows and Eq. 2.1 will no longer apply; that is, the rates of water and solute diffusion will no longer be solely determined by their respective concentration differences. In general, the rate of volume displacement from compartment o to compartment i will now be less than that which can be attributed solely to the flow of solute from compartment i to compartment o.

In the light of these considerations we can redefine the reflection coefficient as follows:

$$\sigma_i = (J_v)_i / J_v \qquad [3.6]$$

where J_v is the osmotic volume flow for a given ΔC_i when *i* is impermeant and $(J_v)_i$ is the volume flow in the presence of the same ΔC_i when the membrane is permeable to i. Thus, when $\sigma_i = 1$, $(J_v)_i = J_v$; when $\sigma_i < 1$, $(J_v)_i < J_v$; and when $\sigma_i = 0$, $(J_v)_i = 0$.

We may now ask, What are the physical determinants of σ_i? Clearly, from the above considerations, if i crosses the membrane by partitioning into and diffusing across the lipid phase, σ_i will be determined by the ability of i to partition into the barrier (K_i), the ease with which it can diffuse by random motion in the lipid (D_i), and the volume occupied by the solute molecule (usually given as the *partial molar volume*, $\bar{v}_i$, i.e., the volume occupied by a mole of i at infinite dilution). Similarly, if i crosses through specific pores that are not the predominant pathways for water flow (i.e., the flows of i and water do not interact), σ_i will be determined by the permeability of the pores to i, the number of such pores per unit area of membrane, and the partial molar volume of i. Referring to Figs. 3.3A and 3.3B, these are the properties that will determine the volume flow of i from compartment i to compartment o and reduce the net displacement of volume due to the oppositely directed flow of water.

If, as illustrated in Fig. 3.3C, solute and water cross the membrane through common pathways, σ_i is determined by the ability of the membrane to "distinguish" between the solute molecule and the water molecule and, for nonelectrolytes, this will be determined by the ratio of the radius of the solute molecule to that of the aqueous pathway or channel. Just as a distant observer cannot distinguish between a canary and an eagle, and a large sewer pipe cannot distinguish between a golf ball and a tennis ball, likewise a pore with a radius of 40 Å would hardly discriminate between a water molecule (radius = 1.5 Å) and sucrose (radius = 6 Å). In an ultrafiltration experiment such as that illustrated in Fig. 3.2, a dilute sucrose solution would pass through a membrane with channels having radii ≈ 40 Å essentially unmodified so that ($C_f \cong C_i$) and $\sigma_{sucrose} \cong 0$. In other words, if, referring to Fig. 3.1, compartment o contains pure water and compartment i contains a 300 mM solution of sucrose and the channels traversing the membrane have radii ≈ 40 Å, $J_v \cong 0$ and $\Delta\pi \cong 0$; under this condition, sucrose is not an osmotically effective solute since the pores cannot distinguish between it and water!

Let us now turn to K_f and, for the sake of simplicity, consider only the case where water or solution flows through pores. It has long been known that fluid flow through tubes whose radii are so large that the fluid can be considered a "continuum" (i.e., the fact that the fluid is made up of discrete molecules that have finite dimensions can be ignored) obeys Poiseuille's law; that is,

$$J_v = (\pi r^4/8\eta\Delta x)\Delta P \qquad [3.7]$$

where r is the radius of the tube, Δx is the length of the tube, and η is the viscosity of the fluid. For reasons that are as yet mysterious, this relation also describes water flow through pores having radii as small as 4 Å so that $K_f = n (\pi r^4/8\eta\Delta x)$, where n is the number of such pores per unit area of membrane.

Finally, again restricting our considerations to porous membranes, we might inquire, Where is the osmotic pressure and how does it arise? As discussed (far more rigorously) by Alexander Mauro and Jack Dainty (see Finkelstein in the bibliogra-

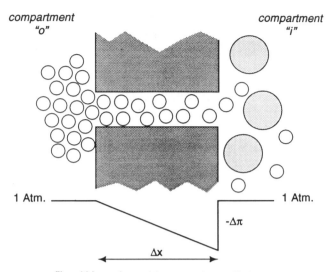

FIG. 3.4. Pressure profile within a channel in a membrane that separates compartment "o," which contains pure water from compartment "i," which contains a solution of an impermeant solute.

phy), the pressure exists within the pore and arises, roughly speaking, from the fact that nature abhors a vacuum! Suppose we could peek into the pore illustrated in Fig. 3.4 and observe the events taking place at the molecular level. As the result of Brownian (random) motion, water molecules will bounce into and out of the pore at both interfaces. But, because the water in compartment i is "diluted" by the presence of an impermeant solute, the likelihood of a water molecule bouncing into the pore across the interface to replace a water molecule that bounced out of the pore into that compartment is reduced; and, the solute is too large to substitute. The result is that local "vacancies" or "rarefactions" arise. In the macroscopic world, rarefaction is associated with a decrease in pressure (recall Boyle's law, which states that the product of the volume of a fixed amount of gas times its pressure is constant at constant temperature; thus, the greater the volume or distance between gas molecules, the lower the pressure the gas exerts on its container). If compartments o and i are open to the atmosphere, there will be a negative (subatmospheric) pressure just within the pore at its interface with compartment i equal to $-\Delta\pi$ and a negative pressure gradient (tension) within the pore between the two interfaces. It is this "invisible" pressure gradient that "drives" the osmotic flow of water through the pore in accordance with Poiseuille's law. If a pressure equal to $\Delta\pi$ is applied to the surface of the membrane facing that compartment (via applying force to a piston as in Fig. 3.1), the gradient within the membrane is abolished and flow ceases.

SOLVENT-DRAG

Before concluding this section on osmosis, let us briefly consider the effect of volume flow on the flow of solute. As discussed above, if water and solute molecules traverse membranes through common pathways, the flows will not be independent but, instead, will interact with or influence one another due to molecular collisions; these interactions between flows may be described using the macroscopic notion of friction. As a result, solute particles can be entrained in the volume flow and carried along much like a leaf in a stream. This phenomenon, referred to as *solvent-drag*, is implicit in the ultrafiltration experiment illustrated in Fig. 3.2.

The important point is that under this circumstance, the flow of a neutral solute is no longer driven by its concentration difference alone as described by Eq. 2.1. Instead, that equation must be expanded to include the effect of solvent-drag as follows:

$$J_i = P_i \, \Delta C_i + (1 - \sigma_i)\dot{C}_i J_v \qquad [3.8]$$

where $\dot{C}_i$ is, roughly speaking, the average concentration of i in the aqueous channel.

We can illustrate this phenomenon with the aid of Fig. 3.1. Let us assume that the membrane separating compartment o from compartment i is traversed by pathways that can accommodate both water and the solute i and that the concentration of i in compartment i is greater than that in compartment o. Now, if $0 < \sigma_i < 1$, there will be a flow of volume from compartment o to compartment i given by Eq. 3.5 (where $\Delta P = 0$). Also, because $C_i^i > C_i^0$, there will be a flow of i in the opposite direction from compartment i to compartment o given by Eq. 3.8. Now, if we define flows from compartment o to compartment i as positive, and flows in the opposite direction as negative, it is clear that J_i will be the difference between the diffusional flow from compartment i to compartment o and the solvent-drag from compartment o to compartment i; the frictional interaction between the oppositely directed flows of water and solute retard the flow of the latter.

Now let us consider another example that is more illustrative of the possible physiological significance of solvent-drag. Assume that the concentrations of solute i in compartments o and i of Fig. 3.1 are equal and that for some reason there is a flow of volume from compartment i to compartment o (e.g., a hydrostatic pressure resulting from the application of force on the piston or an osmotic pressure due to the presence of an osmotically active solute in compartment o alone). Then, according to Equation 3.8, if $\sigma_i < 1$ there will be a flow of i from compartment i to compartment o even though $\Delta C_i = 0$. Indeed, solute i can be "dragged" from compartment i to compartment o because of entrainment in a flow of fluid in that direction even when $C_i^0 > C_i^i$. In other words, coupling of the flow of a solute to the flow of volume (solvent-drag) can result in the movement of a solute against a concentration difference; as will be discussed in Chapter 5, we refer to this uphill movement as *secondary active transport*.

Although solvent-drag is not likely to play a significant role in the movement of solutes across cell membranes, it does contribute to the flow of solutes through large water-filled pathways between endothelial cells that form capillaries and epithelial cells that line organs such as the small intestine.

BIBLIOGRAPHY

Finkelstein A. *Water movement through lipid bilayers, pores, and plasma membranes: Theory and reality*. New York: Wiley-Interscience, 1987.
House CR. *Water transport in cells and tissues*. London: Edward Arnold, 1974.
Schultz SG. *Basic principles of membrane transport*. Cambridge: Cambridge University Press, 1980; Chapter 4.

4

Diffusion of Electrolytes: General Considerations

Chapter 2 deals with the diffusion of uncharged particles, but many of the fundamental properties of the diffusional process described therein also apply to the diffusion of charged particles. In both instances, net flow due to diffusion is the result of random thermal movements, and the diffusion coefficients of the particles are inversely proportional to their molecular or hydrated ionic radii. However, because ions bear a net electrical charge, the diffusion of a salt such as NaCl, which exists in aqueous solution in the form of oppositely charged dissociated ions, is somewhat more complicated for two reasons.

First, because of the attractive force between particles bearing opposite electrical charges, the ions resulting from the dissociation of a salt do not diffuse independently.

Second, under conditions of uniform temperature and pressure, only a difference in concentration can provide the driving force for the diffusion of uncharged particles; net flows do not occur in the absence of concentration differences. For the case of a charged particle, the driving force for diffusion is made up of two components: (1) a difference in concentration; and (2) the presence of an electrical field. The effect of an electrical field on the movement of charged particles is readily illustrated by the familiar phenomenon of electrophoresis. If an anode and a cathode are placed in a beaker that contains a homogeneous solution of NaCl, Cl^- will migrate toward the anode, and Na^+ will migrate toward the cathode. These net movements are due to attractive and repulsive forces that act on charged particles within an electrical field and take place despite the initial absence of a concentration difference.

Each of these aspects of ionic diffusion is separately discussed below.

ORIGIN OF DIFFUSION POTENTIALS

Consider the system illustrated in Fig. 2.1. If a solution containing equal concentrations of urea and sucrose is placed in compartment o and distilled water is placed in compartment i, both urea and sucrose will diffuse across the sintered glass disk independently. Because urea is smaller than sucrose, its diffusion coefficient will be greater than that of sucrose (i.e., $D_{urea} > D_{sucrose}$), and it will move more rapidly in response to the same driving force ($\Delta C_{urea} = \Delta C_{sucrose}$) so that, after sufficient time has elapsed, the solution in compartment i will contain more urea than sucrose. In essence we have employed differential rates of diffusion as a separatory method. If the barrier is permeable to urea but impermeable to sucrose, only urea will appear in compartment i; this is the principle of separation by dialysis. Now, let us replace the solution in compartment o with an aqueous solution of potassium acetate (KAc). This salt dissociates into a relatively small cation (K^+) and a larger anion (Ac^-). If the diffusion of these ions were independent (as in the example with urea and sucrose), it would be possible, in time, to withdraw from compartment i a solution that contains more cation than anion, that is, a solution that has excess K^+. This, in fact, has never been observed. Indeed, the failure to observe a separation of charge in solutions, as well as compelling theoretical arguments, are the bases of a universally accepted postulate known as the *law of electroneutrality*. In essence, this law states that any macroscopic or bulk portion of a solution must contain an equal number of opposite charges; that is, it must be electrically neutral. This is a very important statement that must be kept in mind whenever we consider the bulk movements of ions.

Returning to our example of the diffusion of KAc, we may ask, Why does K^+ not outdistance Ac^-? What is responsible for the maintenance of electroneutrality under these conditions? The answer is that the tendency for K^+ to outdistance Ac^- because of their difference in ionic size (and, hence, diffusion coefficients) is counteracted by the electrostatic attractive forces that exist between these two oppositely charged particles. This mutual attraction, which tends to draw the two ions closer together, exactly balances the effect of the difference in ionic size, which would tend to draw them apart.

One may conceptualize the physical process as follows: In Fig. 4.1 we see the small ion (K^+) and the larger ion (Ac^-) lined up along a partition (the starting line). When the partition is removed, the race begins. Each ion initially moves from left to

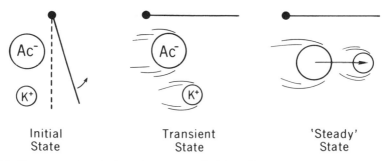

Initial State   Transient State   'Steady' State

FIG. 4.1. Illustration of dipole formation as a result of the differential rates of diffusion of K^+ and Ac^-.

right at the rate that is determined by its ionic size; in other words, each ion initially diffuses from left to right at a rate determined by its "inherent" diffusion coefficient (denoted by D_K and D_{Ac}). Since $D_K > D_{Ac}$, there will be an initial separation of these oppositely charged ions, which, if unopposed, would eventually lead to a violation of the law of electroneutrality. However, the electrostatic attraction between these particles tends to hold them together. Thus, the attraction of Ac^- for K^+ tends to slow down the rate of diffusion of K^+; and, conversely, the attraction of K^+ for Ac^- tends to speed up the rate of diffusion of Ac^-. The net result is that the two ions move from left to right at the same rate but in an oriented fashion (K^+ in front of Ac^-) and thus form a small diffusing dipole (or ion pair). The diffusion coefficient of this ion pair, or dipole, is greater than that of Ac^- alone but less than that of K^+ ($D_K > D_{KAc} > D_{Ac}$). (The situation resembles what would happen if a fast swimmer and a slow swimmer were connected by an elastic cord; there would be an initial rapid separation of the two swimmers, but soon the pair would move together at a rate intermediate to the rate at which each could swim independently.)

Now let us apply this "dipole" concept to diffusion of KAc across a sintered glass disk. Consider the system illustrated in Fig. 4.2. We place a well-stirred solution of

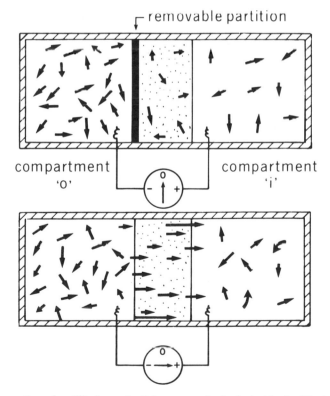

FIG. 4.2. Generation of a diffusion potential as a result of oriented ionic diffusion through a uniform membrane separating two well-stirred compartments.

KAc in compartment o and a more dilute solution of KAc in compartment i; diffusion from o to i is initially prevented by the presence of a solid partition adjacent to the sintered glass disk (Fig. 4.2A). If we now pair up a K^+ ion with a neighboring Ac^- ion so as to form hypothetical dipoles (using the convention that the head of the arrow represents the positive ion), we find, as shown in Fig. 4.2, that the dipoles are randomly oriented simply because the distribution of ions in a homogeneous solution is random. Thus, for every dipole pointed in a given direction there will be another dipole, of equal magnitude, oriented in the exactly opposite direction. The total dipole moments in compartments o and i, as well as within the membrane, are therefore equal to zero, and if electrodes are inserted into the two compartments, the electrical potential difference between these electrodes will be zero.

If we suddenly remove the solid partition, K^+ and Ac^- will diffuse from the higher concentration in compartment o to the lower concentration in compartment i (Fig. 4.2). For the reasons discussed above, this diffusional process can be represented as a series of dipoles crossing the sintered glass disk in an oriented fashion. The random orientations of the dipoles, characteristic of the two well-stirred homogeneous solutions in compartments o and i, have been converted *within* the disk, to an oriented distribution by the presence of a concentration difference *across* the disk. The sum of all these oriented dipoles can be represented by a single dipole whose positive end is pointed toward compartment i and whose negative end is pointed toward compartment o. If electrodes are inserted into compartments o and i, compartment i will be found to be electrically positive with respect to compartment o.

It is important to emphasize that there is *no bulk separation of charges* because the distance between the leading K^+ ion and the lagging Ac^- ion averages only a few (~ 10) angstrom units. The electrical potential difference across the membrane is not due to a bulk (chemically detectable) separation of charge but to the fact that the ions cross the disk in an oriented fashion rather than in a random fashion. In essence, the orientation of the dipoles within the disk can be viewed as having converted the disk into a battery with the positive pole facing compartment i.[1]

The electrical potential difference arising from the diffusion of ions derived from a dissociable salt from a region of higher concentration to one of lower concentration is referred to as a *diffusion potential*. It arises whenever the ions resulting from the dissociation of the salt differ with respect to their mobilities or diffusion coefficients. The orientation of the diffusion potential is such as to retard the diffusion of the ion having the greater mobility and to accelerate the diffusion of the ion with the lower mobility so as to maintain electroneutrality. Thus, the magnitude of a diffu-

---

[1]Technically speaking, there is a small, initial separation of charges sufficient to charge the capacitance of the membrane. This is completed very rapidly and thereafter the movements of anions and cations across the membrane is one-to-one. Furthermore, the amount of charge separation is minute compared to the number of ions present in the two solutions so that bulk electroneutrality is preserved (see also Chapter 9).

sion potential will be directly dependent on the difference between the mobilities of the anion and cation.

It can be shown that the diffusion potential (V) arising from the diffusion of a salt that dissociates into a monovalent anion and a monovalent cation is given by

$$V = \left\{\frac{D_+ - D_-}{D_+ + D_-}\right\} \times \left\{\frac{2.3RT}{F}\right\} \times \log\left\{\frac{C_i^o}{C_i^i}\right\} \qquad [4.1]$$

where D_+ is the diffusion coefficient of the monovalent cation; D_- is the diffusion coefficient of the monovalent anion; C_i^o and C_i^i are the concentrations of the salt in compartments o and i, respectively; R is the gas constant; T is the absolute temperature; and F is the Faraday constant. At 37°C ($T = 310°K$), $2.3\,RT/F = 60\text{mV}$.

Thus,

$$V = \left\{\frac{D_+ - D_-}{D_+ + D_-}\right\} \times 60 \log\left\{\frac{C_i^o}{C_i^i}\right\} \text{ (mV)} \qquad [4.2]$$

Equation 4.2 discloses the following characteristics of diffusion potentials:

$V = 0$ When $C_i^o = C_i^i$

Clearly, when $C_i^o = C_i^i$, there can be no net flow due to diffusion. Diffusion potentials can only arise in a system containing ion gradients.

$V = 0$ When $D_+ = D_-$

Clearly, if both ions have equal mobilities (or sizes), there is no "inherent leader" and no "inherent lagger" so that dipoles are not formed. Another way of viewing this condition is that when both ions have the same mobility, a one-to-one flow across the disk (i.e., bulk electroneutrality) is assured by the fact that both ions are driven by the same ΔC; a diffusion potential is not "necessary" for the preservation of bulk electroneutrality.

When $C_i^o \neq C_i^i$

When $C_i^o \neq C_i^i$, the magnitude of V is directly proportional to the difference between the individual ionic diffusion coefficients. The orientation of V is such that it retards the more mobile ion and accelerates the less mobile ion. Thus, if the mobility of the anion exceeds that of the cation, the more dilute solution will be electrically negative with respect to the more concentrated solution. Conversely, if the cation has a greater mobility than the anion, the more dilute solution will be electrically positive with respect to the concentrated solution.

A particularly interesting situation arises when the barrier is impermeant to one of the ionic species, say, for example, the cation. When $D_+ = 0$, Eq. 4.2 reduces to

$$V = -60 \log\left\{\frac{C_i^o}{C_i^i}\right\} \text{ (mV)} \qquad [4.3]$$

The same expression, with the opposite sign, is obtained if we set $D_- = 0$.

Equation 4.3 states that, when one of the ionic species of a dissociated salt cannot penetrate the barrier, V is dependent only on the concentration ratio across the membrane and is independent of the permeability (or diffusion coefficient) of the ion that can penetrate the barrier. The reason for this independence becomes evident when we recall the law of electroneutrality. If the membrane is impermeant to one of the ions, there can be no net flow of the other, permeant, ion across the membrane, or bulk electroneutrality would be violated. Thus, despite the presence of a concentration difference, there can be no net diffusion of salt across the barrier, and the system is in a state of equilibrium. Because the net flow of the permeant ion is prohibited, the mobility of this ion is of no importance.

The following example may serve to clarify this point and the roles played by concentration differences and electrical potential differences in the overall driving force for the diffusion of ions. Let us place a 0.1-M solution of K^+-proteinate in compartment o and a 0.01-M solution of the same salt in compartment i. If the barrier is impermeant to the large proteinate anion, the electrical potential difference across the barrier at 37°C will be

$$V = 60 \log\{0.1/0.01\} = 60 \text{ mV}$$

with the dilute solution electrically positive compared to the concentrated solution. Now we may ask, Why is there no diffusion of K^+ from compartment o to compartment i down a tenfold concentration difference despite the fact that the membrane is highly permeable to K^+? The answer derives from the fact that there are two forces acting on the K^+ ion. There is a *chemical* force arising from the fact that the concentration of K^+ in compartment o is ten times that in compartment i; this force tends to drive K^+ from compartment o to compartment i. In addition, Eq. 4.3 tells us that there is a 60-mV electrical potential difference between compartments o and i, with compartment i electrically positive. This electrical potential difference tends to drive the positively charged K^+ ion from compartment i to compartment o. These two oppositely directed driving forces (the *chemical* force and the *electrical* force) exactly balance each other so that there is no net driving force for the diffusion of K^+ and, hence, no net movement.

Equation 4.3 was derived by the great German physical chemist Walther Hermann Nernst (1864–1941) from thermodynamic considerations and is referred to as the *Nernst equation* (also under these conditions it is often referred to as the *Nernst potential* or the *Nernst equilibrium potential*). In essence, it embodies the fact that (under conditions of uniform temperature and pressure) there are only two driving forces that influence the diffusion of charged particles; a force arising from concentration differences and a force arising from electrical potential differences. When solutions having different concentrations of the same dissociable salt are placed on opposite sides of a barrier that is impermeant to one of the dissociation products, there will be no net movement of salt across the barrier despite the concentration difference. Net movement of the permeant ion is prevented by the development of an electrical potential difference across the barrier whose magnitude and orientation are such that they exactly cancel the driving force arising from the chemical concentration difference across the barrier. [A more detailed discussion of this point is beyond the scope of this presentation. In essence, the two forces that influence the movements of charged particles—concentration differences and electrical potential differences—can be converted into the same units of force per ion or force per mole (e.g., dynes/mol), and a tenfold concentration ratio at 37°C exerts the same force on a monovalent ion as does a 60-mV electrical potential difference. When, as in the above example, the two forces are oriented in opposite directions, the net force is zero, and there can be no net flow].

CRITERIA FOR ACTIVE TRANSPORT OF IONS

In Chapter 2, we distinguished between passive (or downhill) and active (or uphill) transport of *uncharged* molecules. The sole criterion for this distinction is the relation between the direction of the net movement of the molecule and the direction of the concentration gradient. If an *uncharged* molecule moves or is transported from a region of higher concentration to one of lower concentration, the transport process is said to be passive (or downhill) because the flow is in the direction of the driving force and thus can be attributed entirely to the thermal energy prevalent at any ambient temperature. Conversely, if the molecule moves or is transported from a region of lower concentration to one of higher concentration, the flow is termed active (or uphill) because thermal energy alone cannot account for this movement, and additional forces must be involved.

The distinction between active and passive transport of ions is slightly more complicated because both concentration differences and electrical potential differences can provide driving forces for the diffusional movements of charged particles. This important point may be clarified by the following example.

Consider the movement of K^+ across a membrane separating compartment o from compartment i (see Fig. 2.1) under the following conditions: (1) the concentration of K^+ in compartment o is 0.1 M; (2) the concentration of K^+ in compartment i is 0.2 M; and (3) compartment i is electrically negative with respect to compartment o by 60 mV. Thus, the chemical force acting on K^+ is a twofold concentration ratio tending to drive K^+ from compartment i to compartment o. The electrical force acting on K^+ is a 60-mV electrical potential difference tending to drive it from compartment o to compartment i. As we noted previously, a 60-mV electrical potential difference is equivalent to a tenfold concentration ratio; thus, in this example, the electrical driving force exceeds the chemical driving force, and K^+ will diffuse from compartment o to compartment i spontaneously, even though the direction of net movement is against the concentration gradient.

Now, in general, we can determine the equivalent electrical force, E_i, corresponding to a given concentration ratio by using the Nernst equation, that is,

$$E_i = (RT/zF) \ln(C_i^o/C_i^i) \qquad [4.4]$$

where z is the valence of the ion i; and E_i is the equivalent electrical potential of compartment i minus that of compartment o. Thus, if compartment o is arbitrarily chosen as the ground state and is defined as having a zero electrical potential,[2] then E_i is the equivalent electrical potential difference corresponding to a given concentration ratio across the membrane with the magnitude and sign of compartment i.

At 37°C, Eq. 4.4 reduces to

$$E_i = (60/z) \log(C_i^o/C_i^i) \text{ (mV)} \qquad [4.5]$$

Now, using Eq. 4.5, we can determine the direction in which an ion will diffuse in the artificial system illustrated in Fig. 2.1, knowing the concentrations of the ion in compartments o and i and the electrical potential difference across the membrane V_m, where V_m is the electrical potential of compartment i with respect to that of compartment o.

Let us illustrate this point by considering some examples.

Example 1. $C_i^o = 10$ mM; $C_i^i = 100$ mM; $z = +1$; and $V_m = -60$ mV.

Substituting these values into Eq. 4.5, we obtain $E_i = 60 \log (1/10) = -60 \log(10/1) = -60$ mV. Thus, the force of the tenfold concentration ratio which is driving the cation to flow from compartment i to compartment o is equivalent to 60 mV. But at the same time, compartment i is 60 mV *negative* with respect to compartment o and this force tends to "hold" the cation in compartment i. Thus, the net force on the cation is zero, and there will be no net movement across the membrane in either direction. The system is at equilibrium.

In short, when $E_i = V_m$, the chemical driving force arising from the concentration difference (ratio) across the membrane is exactly counterbalanced by the electrical driving force across the membrane, and there will be no flow.

Example 2. $C_i^o = 5$ mM; $C_i^i = 100$ mM; $z = +1$; and $V_m = -60$ mV.

Now, from Eq. 4.5, $E_i = -78$ mV. Thus, the chemical driving force for the flow of i from compartment i to compartment o is 78 mV, but the electrical holding force is only 60 mV. Thus, i will diffuse across the membrane from compartment i to compartment o.

Example 3. $C_i^o = 20$ mM; $C_i^i = 100$ mM; $z = +1$, and mV $= -60$ mM.

---

[2] A voltmeter does not measure an electrical potential but rather the difference between the electrical potentials of the two points to which its electrodes are connected. Thus, when we refer to the electrical potential of one compartment, it must be with reference to that of the other compartment, which is chosen as the "ground" or zero potential compartment. The universally accepted convention today is that the extracellular (or outer, o, compartment) is designated to be the ground or reference in electrophysiological studies.

In this case, $E_i = -42$ mV. Thus, the chemical driving force tending to "push" i from compartment i to compartment o is less than the electrical driving force tending to "pull" the cation from compartment o to compartment i. The net result will be the diffusion of i from compartment o into compartment i despite the fact that this flow is against a concentration difference.

Summary

An uncharged substance is said to be actively transported if net movement is directed against a concentration difference. An ion is said to be actively transported only if its net movement is directed against a *combined* concentration and electrical potential difference; flow of an ion from a region of lower concentration to one of higher concentration is not by itself inconsistent with simple diffusion. The Nernst equation permits us to determine whether the movement of a charged solute is passive or active.

For an uncharged solute, the rate and direction of diffusion across a membrane can be described by the equation

$$J_i = P_i(C_i^o - C_i^i) \qquad [4.6]$$

where we define J_i as positive when the flow takes place from compartment o to compartment i; thus, when $C_i^i > C_i^o$, the flow is from compartment i to compartment o and is negative.

For an ion, J_i is determined by the permeability of the membrane and the total electrochemical driving force. Now, it is often convenient to express J_i as a current given by $I_i = z_i F J_i$, where z_i is the valence of the ion, and F is the Faraday constant, which has the value of 96,500 coulombs/mol. Thus, if J_i is in units of moles per unit area per unit time (e.g., mole/cm^{2}hr), then $I_i = z_i F J_i$ is in units of coulombs/cm^2 hr. Because electrical current flow is given in amperes, which is in units of coulombs/sec (i.e., the amount of charge flowing per second), I_i is in units of amperes per unit membrane area, or A/cm^2.

Now, the equation describing the rate of diffusion of an ion is

$$z_i F J_i = I_i = G_i(V - E_i) = G_i \left[V - \frac{2.3 \, RT}{z_i F} \log \frac{C_i^o}{C_i^i} \right] \qquad [4.7]$$

where the total *electrochemical* driving force is $V - E_i$ (analogous to ΔC_i for an uncharged particle) and takes into account the two forces acting on charged particles; and G_i is the conductance of the membrane to i (the inverse of the resistance of the membrane to the flow of i) and is analogous to P_i.

Equation 4.7 has the form of Ohm's law (i.e., $I = GV$), which, as discussed above, has the same form as Fick's (first) law of diffusion, which relates the rate of diffusion of an uncharged solute to its driving force.

BIBLIOGRAPHY

Hille B. *Ionic channels of excitable membranes*. Sunderland, MA: Sinauer Associates, Second Edition, 1992.

Schultz SG. *Basic principles of membrane transport*. Cambridge: Cambridge University Press, 1980; Chapters 2–3.

5

Ion Diffusion through
Biological Channels

In Chapter 4, we considered the diffusion of ions across barriers without specifying the nature of the pathway(s) traversed. Because all of the reasoning was based on thermodynamic principles, the conclusions arrived at in that chapter are valid regardless of the nature of the barrier; that is, it could be a sintered glass disk, a sheet of cellophane, a sheet of filter paper, a lipid bilayer—whatever! Turning now to biological membranes, it has long been recognized that because ions are scarcely soluble in lipids, they can only cross those barriers by reversibly combining with some mobile component of the membrane and/or by diffusion through aqueous channels. There is now undeniable evidence for both types of mechanisms and, in this chapter, we will focus on the latter.

Let us start by inquiring how rapidly ions might be expected to traverse a lipid bilayer through an aqueous pathway assuming that the diffusion coefficient is the same as that in free aqueous solution. Let us assume that the ion has a "naked" or crystal radius, $r_i = 0.15$ nm, and that its diffusion coefficient $D_i = 2 \times 10^{-5}$ cm^2/sec; these are close to the values for K^+. Let us further assume that the channel has a radius, r_p, of only 0.2 nm, that the bilayer has a thickness, $\Delta x = 5$ nm, and that the concentration difference across the channel $\Delta C_i = 100$ mM. From Eq. 2.1 it follows that

$$J_i = \left(\frac{\pi r_p^2 D_i \Delta C_i}{\Delta x}\right) = 4 \times 10^{-18} \text{ mol/sec}$$

and, multiplying by Avagadro's number, $J_i = 2.5 \times 10^6$ ions/sec. If the ion is monovalent, then, since the electronic charge is $e = 1.6 \times 10^{-19}$ coulombs the current attributable to this flux of ions is 0.4×10^{-12} coulombs/sec or 0.4 pA. The important point of this exercise is that "ballpark" estimates are consistent with the notion that more than a million ions might be able to diffuse per second across a lipid bilayer through a very "snug" pore and that this would give rise to a current that can be readily measured using available amplifiers.

Since the mid-1970s, measurements of bursts of ionic currents across biological membranes have proven to be entirely consistent with the above estimate and, together with other evidence, leave little doubt that these bursts are the results of ion movements through pores or channels that are integral membrane proteins.

The two methods that are employed to measure single-channel activities are schematized in Fig. 5.1. The first, (A), was introduced in 1976 and is referred to as the "patch-clamp" technique. Briefly, a polished glass micropipette tip having a diameter of approximately 1 micron is pressed against a patch of cell membrane forming a seal that is essentially "leak proof" to ions (seal resistance is greater than 1 billion ohms). Thus, if there happens to be an ion channel in the membrane patch, ionic currents into or out of the cell will be entirely constrained within the pipette and the connected circuitry. One may also mechanically rip (excise) the patch of membrane together with the channel off the cell and examine single-channel properties under artificial but well-defined and easily controlled conditions.

The other technique involves reconstituting or incorporating membrane vesicles containing channels, or purified channel proteins, into planar artificial lipid bilayers separating two easily accessible and controlled solutions (Fig. 5.1B).

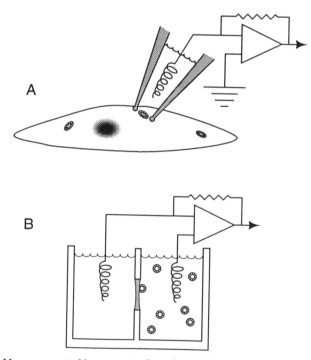

FIG. 5.1. (A) Measurement of ion currents through a single channel by "clamping" a patch of membrane with a micropipette. (B) Apparatus for reconstituting vesicles containing channels into a planar lipid bilayer. The current–voltage (i–V) converter measures the small membrane currents by determining the voltage developed across a very large resistor.

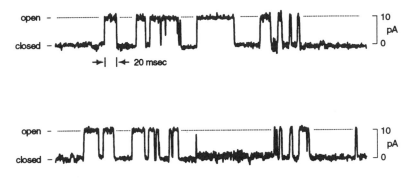

FIG. 5.2. Bursts of current accompanying the random openings of an ion channel.

A recording of the activity of a single ion channel is illustrated in Fig. 5.2. [Note that the ordinate is in pA (i.e., 10^{-12} A) and the abscissa is in milliseconds.] The openings are characterized by abrupt upward deflections from the baseline that reaches a plateau and then abruptly close. An important characteristic of single-channel activity is that its openings and closings are stochastic or random processes. That is, when a channel will open and how long it will remain open under fixed conditions are independent of its past history (e.g., how long it was closed before this opening, how long it was open the last time it was open); the channel does not have a memory! This characteristic enables one to analyze single-channel activity using statistical or probability theory. For example, by analyzing a large number of opening and closing events such as those shown in Fig. 5.2, one can determine the "open-time" probability of the channel, P_o; that is, the likelihood that the channel will be found in the open or conducting state at any given moment or the fraction of time the channel is open. Furthermore, if the channel can only randomly switch between two states due to thermal motion then the transition between these states is given by

$$\text{Closed state} \underset{k_{-1}}{\overset{k_1}{\rightleftharpoons}} \text{Open state}$$

where the ks are rate constants analogous to those employed in chemical kinetics. In this instance, k_1 is the probability of transitions from the closed to the open state in unit time so that $(1/k_1)$ is the mean duration of the closed states or the mean closed time (τ_c). Likewise, k_{-1} is the probability of transitions from the open state to the closed state in unit time, so that $(1/k_{-1})$ is the mean open time (τ_o). It follows that the open-time probability will equal the mean open time divided by the "mean total" time; hence, $P_o = \tau_o/(\tau_o + \tau_c)$.

Now, because of the random nature of a channel's openings and closings, the mean duration of either the open or closed states is determined only by the rate constants acting *away* from that state. Thus, by analogy with a simple first-order chemical reaction, we can write $d(\text{open})/dt = -k_{-1}(\text{open})$, where the term in parentheses, roughly speaking, refers to the duration of a single channel in the open state. Integrating this equation we obtain

$$(\text{open})_t = (\text{open})_o e^{-tk_{-1}} = (\text{open})_o e^{-t/\tau_0} \qquad [5.1]$$

which states that once a channel opens the likelihood or probability that it will remain open decays exponentially with a rate determined by k_{-1} (or τ_o); the probability that a channel, once opened, will remain open for t msec or longer is given by e^{-t/τ_o}.

We can determine the mean open time (τ_o) by examining a large number of open events of a single channel and sorting them into "bins" based on their durations. For example, the first bin could contain the number of events when the channel was open for at least 0.5 msec; the second bin could contain all of the open events lasting at least 1 msec; the third, all of the open events lasting at least 1.5 msec; and so on in 0.5 msec increments. We can then construct a histogram as shown in Fig. 5.3 where the ordinate is the number or frequency of events per 0.5 msec and the abscissa is open time (t, in msec). If the channel conforms to the two-state model (i.e., it is either fully closed or fully open) then the midpoints of the histogram bars

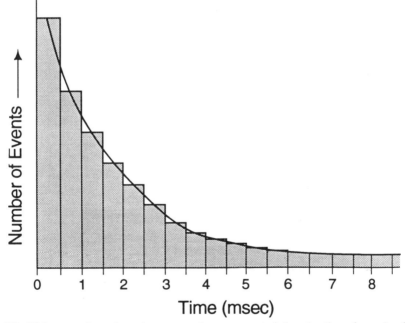

FIG. 5.3. Histogram of number of open events versus cumulative duration of opening for a single channel observed over a long period of time. The curve corresponds to an exponential decay where $\tau_o \approx 2$ msec. Thus ~63% of the openings had durations less than 2 msec and ~37% had longer open times.

can be joined by a curve described by a single exponential decay with increasing time and k_{-1} or τ_o can be derived by simple curve-fitting procedures. A similar approach can be employed to determine the mean closed time, τ_c. If, however, the channel's behavior is more complicated, the open-time and closed-time histograms will not conform to single exponentials. For example, a channel could reside in one of three possible states—very closed, not quite as closed but still not open, and open. In this case, one exponential might describe the open-time probability histogram but two might be needed to describe the closed-time histogram.

Finally, by determining single-channel activity when the membrane is clamped at a number of different electrical potential differences one can obtain valuable information regarding the conductance and ionic selectivity of the channel. For example, Fig. 5.4 illustrates the relation between the size of ionic currents flowing through a single channel (i_c) and the membrane potential, V_m, when the solution facing the outer surface of the channel contained 15 mM KCl and that facing the inner surface contains 150 mM KCl. Note that in this example, the relation between i_c and V_m is linear (or Ohmic) and can be described by a relation analogous to Eq. 4.7

$$i_c = g_c(V_m - V_r) \qquad [5.2]$$

where g_c, the slope, is the conductance of the single channel. [Recall that the accepted convention is that the outer or extracellular solution is considered "ground" (zero potential) so that V_m is the electrical potential of the inner compartment with respect to that in the outer compartment and that the flow of cations (i.e., a positive current) from the inner or intracellular compartment to the outer compartment results in an upward or positive deflection.]

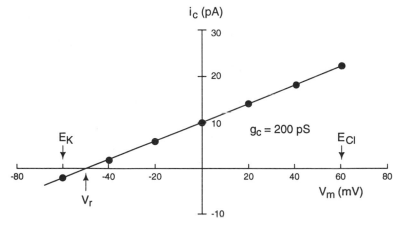

FIG. 5.4. Relationship between single channel current, i_c, and the electrical potential difference across the channel, V_m, in the presence of asymmetric KCl solutions. E_K and E_{Cl} are the Nernst equilibrium potentials for K^+ and Cl^- and V_r is the electrical potential difference across the channel when current flow is zero (i.e., the "reversal" or "equilibrium" potential for the channel under the given conditions). g_c is the conductance of the channel.

Now, using the Nernst equation (4.5) we can calculate the equilibrium potentials for K^+ (i.e., E_K) and Cl^- (i.e., E_{Cl}); recall that E_K is the value of V_m at which there is no ionic flow if the membrane were permeable to K^+ but impermeable to Cl^-, and E_{Cl} is the value of V_m at which there is no ionic flow if the membrane were permeable to Cl^- and impermeable to K^+. As indicated on Fig. 5.4, for this tenfold concentration ratio, $E_K = -60$ mV and $E_{Cl} = 60$ mV. Note, however, that the observed value of V_m at which the current is zero, which is referred to as the *zero-current* or *reversal* potential, V_r, is -50 mV. The fact that V_r is not equal to either E_K or E_{Cl} indicates that the channel is not exclusively permeable to either K^+ or Cl^-. Moreover, the observation that V_r is much closer to E_K than to E_{Cl} indicates that the channel is much more permeable to K^+ than to Cl^-. The actual ratio of P_K/P_{Cl} can be obtained using the expression for the *diffusion potential* (Eq. 4.2), which can be written as follows:

$$V_r = \left\{ \frac{(P_K/P_{Cl}) - 1}{(P_K/P_{Cl}) + 1} \right\} 60 \log \left\{ \frac{(KCl)_o}{(KCl)_i} \right\} \qquad [5.3]$$

Solving this equation for the condition, $(KCl)_o = 15$mM, $(KCl)_i = 150$mM, and $V_r = -50$ mV yields $(P_K/P_{Cl}) = 11$.

Clearly, if the channel were ideally permselective for K^+, the current–voltage relationship shown in Fig. 5.4 would have been described by the equation $i_K = g_K (V_m - E_K)$ and the intercept on the abscissa would be where $V_m = -60$ mV.

DETERMINANTS OF CHANNEL SELECTIVITY

One amazing property of many biological ion channels is their ability to sharply distinguish among ions of the same charge whose dimensions differ by less than 0.1 nm. For example, as we will consider in greater detail below, some channels in nerve and muscle membranes may be 100 times more permeable to K^+ than to Na^+ in spite of the fact that the former has a radius of 0.133 nm, whereas the latter is *actually smaller* and has a radius of 0.095 nm!

The best explanation for the exquisite ability of many channels to discriminate among ions with a high degree of selectivity is the *closest fit* theory advanced by Hille. It has long been recognized that because water is a polar molecule, ions float around in aqueous solution in association with a cloud or shell of water molecules; that is, ions in aqueous solution are hydrated. Furthermore, the electrostatic attractions between ions and water molecules are quite strong (recall that the heat of solution of salts can be quite large) so that a considerable amount of energy is needed to dehydrate an already hydrated ion. Now, Hille argues, suppose the steric arrangement of fixed charges in the region of the channel that determines ionic selectivity is such that an ion traversing that region can be stripped of its water of hydration without its knowing it! In other words, if, as illustrated in Fig. 5.5 for the case of a cation, the water of hydration can be replaced by the negative poles of amino acids that line the channel (e.g., carboxylate groups) in such a way that the

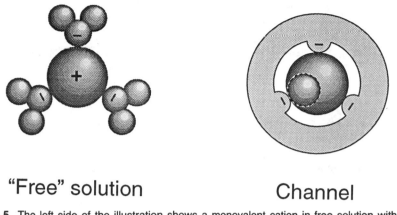

"Free" solution   Channel

FIG. 5.5. The left side of the illustration shows a monovalent cation in free solution with the hydration shell of three water molecules. On the right, the same cation is in a cylindrical channel where the interaction with electronegative fixed charges in the wall of the pore exactly mimics the interactions with the water dipoles. The dashed circle represents a smaller cation that cannot fit in the channel while retaining its hydration shell but is energetically uncomfortable without it.

cation is as "energetically comfortable" in the channel as it is in water, then the ion would not recognize the fact that it left its aqueous environment for that of the channel. (In the language of thermodynamics, the energy needed to partition into or out of the channel would be negligible so that according to the Boltzman distribution, the probability of being in the channel is equal to the probability of being in the aqueous solution; this is analogous to an amphipathic molecule whose oil–water partition coefficient is unity so that it is just as comfortable in aqueous solution as in a lipid solvent.) Clearly, if the radius of the dehydrated ion is too large to be accommodated by the channel, it will be excluded. But, if the radius of the dehydrated ion is too small to fit "snugly" (dashed circle in Fig. 5.5), it will be energetically disadvantaged compared to the ion with the closer fit; it will not be as willing to shed its comfortable coat of water molecules for the more foreign environment of the channel, and its partitioning into the channel will be energetically less favorable than that of the ion with the closer fit.

REGULATION OF CHANNEL AND MEMBRANE CURRENTS

Up to this point, we have considered single ion channels whose currents are given by an equation having the form of Eq. 5.2. We also observed that a single channel is not open all of the time, but, instead, undergoes spontaneous transitions between open and closed states. It follows that the total current of an ion, i, across a unit area of membrane containing an ensemble of channels selective for i that are opening and closing at random, is given by

$$I_i = N_i P_o i_i = N_i P_o g_i (V_m - E_i) \qquad [5.4]$$

where N_i is the total number of single channels per unit membrane area and P_o is the open-time probability. Comparing Eq. 5.4 with Eq. 4.7, we see that the total conductance of a membrane to i, G_i, is simple $N_iP_og_i$. Furthermore, because (as we will soon see) P_o and g_i may be functions of V_m, a more general expression of Eq. 5.4 is

$$I_i = N_iP_{o(Vm)}i_i = N_iP_og_{i(Vm)}[V_m - E_i] \qquad [5.5]$$

Now, under steady-state conditions when C_i^o and C_i^i are constant so that E_i is constant (see Eq. 4.7), changes in I_i can result only from changes in N_i, P_o, and/or g_i. Under conditions where N_i is constant, physiological regulation of channel activity or I_i is the result of changes in P_o; that is, the channel is either nonconducting or fully conducting at fixed values of V_m and E_i and the parameter that is regulated is the fraction of time that the channel is in either of these two states.

The two major physiological determinants of P_o are V_m and/or chemical regulators. Thus, channels are roughly categorized as "voltage-gated" or "ligand-gated." It should be emphasized, however, that this classification is not ironclad; many channels that are predominantly considered voltage-gated are also influenced by chemical regulators and some ligand-gated channels are influenced by V_m.

BIBLIOGRAPHY

Cook NS (Ed.). *Potassium channels: structure, classification, function and therapeutic potential.* Chichester, England: Ellis Horwood Ltd., 1990.

Hille B. *Ionic channels of excitable membranes*, 2nd ed. Sunderland, MA: Sinauer Associates Inc., 1992.

Miller C (Ed.). *Ion channel reconstitution.* New York: Plenum Press, 1986.

Sakmann B, Neher E (Eds.). *Single-channel recording.* New York: Plenum Press, 1983.

6

Carrier-Mediated Transport

During the first quarter of this century, compelling evidence began to surface that life would not be possible if diffusion were the only mechanism available for the exchange of solutes across cell boundaries. There are two sets of observations that most strongly implied the necessity for additional transport mechanisms:

1. Most biological membranes are virtually impermeable to hydrophilic molecules having molecular radii significantly greater than 4 Å or that have five or more carbon atoms. Thus, virtually all essential nutrients and building blocks (e.g., glucose, amino acids) cannot penetrate biological membranes to any significant extent by diffusion so that other mechanisms are necessary to provide for their entry into cells. Similarly, biological membranes are generally impermeant to essential multivalent ions such as phosphate so that their movements across cell membranes must also be mediated by mechanisms other than diffusion.

2. The intracellular concentrations of many water soluble solutes differ markedly from their concentrations in the extracellular medium bathing the cells. For example, a characteristic of virtually every cell in the animal and plant kingdoms is that the intracellular K^+ concentration greatly exceeds that in the extracellular fluid (in some cases by a factor of more than 1,000 to 1), and in cells from higher animals the intracellular Na^+ concentration is much less than that in the bathing media (often by a factor greater than 10). This ionic asymmetry is important for a number of vital processes and, as we shall see, is the basis of many bioelectric phenomena that play an essential role in nerve conduction and muscle contraction. Diffusional processes alone cannot be responsible for the production and maintenance of these asymmetries.

To accommodate these two sets of observations, a concept referred to as *carrier-mediated transport* or the *carrier hypothesis* evolved. This hypothesis is generally attributed to Osterhout, who, in 1933, suggested that biological membranes contain components ("carriers") that are capable of binding a solute molecule at one side of the membrane to form a carrier-solute complex, which then crosses the membrane, dissociates, and discharges the transported solute on the other side.

Since then an overwhelming body of evidence for the role of membrane compo-

45

nents, or carriers, in biological transport processes has accumulated. Carriers have been implicated in the transport of a wide variety of solutes, and the specific properties of numerous carrier systems have been described in considerable detail. Needless to say, a comprehensive discussion of biological carriers is beyond the scope of this presentation, and we will limit ourselves to a brief consideration of some of the general characteristics of carrier-mediated transport processes.

SOME GENERAL CHARACTERISTICS OF CARRIER-MEDIATED TRANSPORT

The following features are so widely characteristic of carrier-mediated transport processes that they are generally considered sufficient and, often, necessary criteria for the implication of carriers in the transport of a given solute:

1. Virtually all carriers appear to display a high degree of structural specificity with regard to the substances they will bind and transport. For example, the carriers responsible for the transport of glucose into animal cells are highly stereospecific; they will rapidly bind and transport the dextrorotary form (D-glucose) but have little affinity for the levorotary form (L-glucose). Conversely, the carriers responsible for the transport of amino acids into animal cells possess a high degree of selectivity in favor of the L-stereoisomer and little affinity for the D-stereoisomer.

2. All carrier-mediated transport processes exhibit *saturation kinetics*; that is, the rate of transport gradually approaches a maximum as the concentration of the solute transported by the carrier increases. Once this maximum rate is achieved, a further increase in the solute concentration has no effect on the transport rate. Plots of the rate of transport against concentration often closely resemble the hyperbolic plots characteristic of Michaelis-Menten enzyme kinetics, and, under these conditions, the kinetics of the transport process can be described by defining the maximum transport rate (J_{max}) and the substrate concentration at which the transport rate is half-maximum (K_t). Thus, $J_i = [J_{max}(C_i)/(K_t + (C_i))]$, where C_i is the concentration of the solute.

A graph illustrating the saturation kinetics characteristic of carrier-mediated transport is shown in Fig. 6.1B. In contrast, as illustrated by Fig. 6.1A, transport due to simple diffusion is usually characterized by a linear relation between transport rate and solute concentration as predicted by Eq. 2.4. It should be emphasized that ionic diffusion through channels may exhibit saturation but, usually, only when concentrations are well beyond the physiological range.

The saturation phenomena observed in carrier-mediated transport processes reflect the presence of a fixed and limited number of carrier molecules or binding sites in the membrane. When the solute concentration is sufficiently high so that all of the carrier sites are occupied (or complexed), a further increase in concentration cannot elicit a further increase in transport rate.

One consequence of the presence of a limited number of carrier molecules for a given class of transported solutes is the phenomenon of competitive inhibition. This

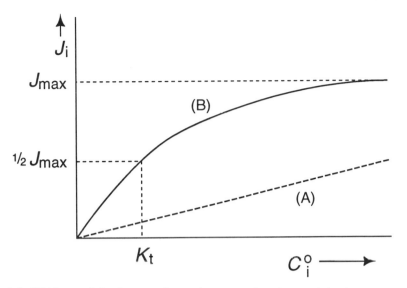

FIG. 6.1. **(A)** Linear relation between flux and concentration, characteristic of many diffusional processes. **(B)** Relation between flux J_i and concentration C_i for a saturable process illustrating J_{max} and K_t (the concentration at which J_i is one-half J_{max}).

is observed when two or more solutes that are capable of being transported by the same carrier are present simultaneously, competing with one another for the limited number of available binding sites. This phenomenon is closely analogous to competitive inhibition in enzyme–substrate interactions and often may be also described by classical Michaelis-Menten kinetics.

In addition to providing evidence for carrier-mediated transport, the phenomenon of competitive inhibition has proved to be extremely useful for the purpose of defining the transport specificity of a given carrier mechanism. Thus, if two solutes A and B are each transported by carriers and exhibit mutual, classical *competitive inhibition*, one may conclude that the same carrier mechanism is involved; if they in no way compete with each other, at least two distinct carrier mechanisms must be involved. For example, all of the D-hexoses that are absorbed by the small intestine mutually compete with one another for the same limited transport system. When glucose and galactose are separately present in the intestinal lumen at high concentrations, they are each absorbed at approximately the same maximum rate. On the other hand, if the same concentrations of glucose and galactose are instilled into the intestinal lumen in the form of a mixture, each will be absorbed at a rate significantly lower than that observed when they were present separately. The total rate of sugar transport will be equal to the maximum rates observed when each sugar was present separately, indicating that the two sugars are competing for the same carrier system and are sharing in the saturation of the total number of available sites. On the other hand, the transport of glucose is not inhibited by the presence of hexoses that are not subject to carrier-mediated transport.

NATURE OF MEMBRANE CARRIERS

The characteristics of carrier-mediated transport processes that we have just described, namely, (1) the high degree of structural specificity and (2) saturation kinetics and competitive inhibition, strongly resemble the characteristics of enzyme–substrate interactions. After the introduction of the carrier hypothesis, it was suspected that carriers were enzymelike molecules that comprise part of the protein portion of the lipoprotein membrane. However, for many years, these carriers defied isolation and characterization, and until recently there was a relatively large group of investigators who doubted, and even denied, their existence; but the results of numerous studies during the past few decades have dispelled these doubts. The development of techniques for isolating cell membranes and gently detaching their protein components has led to the isolation of integral proteins that are capable of specifically binding transported solutes. In many instances, these purified proteins have been reinserted into artificial lipid membranes, and these artificial (reconstituted) systems are capable of mediating the transport of specific solutes.

In short, considerable progress has been made toward defining the biochemical and/or molecular basis of carrier-mediated transport. The precise mechanism(s) by which the transported solutes are translocated across the membrane after binding, however, remains a mystery. But, in light of our current understanding of the assembly of proteins in biological membranes, it is certain that the notions that carriers are "ferry boats" or that integral proteins "flip-flop" across the lipid bilayer are incorrect. It is more likely that carriers are integral protein that in many respects resemble channels and that binding and translocation of solutes from one side of a "gate" to the other takes place within these channels.

FACILITATED DIFFUSION AND ACTIVE TRANSPORT

As discussed above, the two functions that membrane carriers must fulfill are (1) to provide a mechanism by which otherwise impermeant solutes can enter or leave cells across membranes and (2) to provide a mechanism by which substances can be actively transported into or out of cells. The two classes of carrier-mediated transport processes that fulfill these functions are referred to as *facilitated diffusion* and *active transport*, respectively.

Facilitated Diffusion

Facilitated diffusion is the term reserved for carrier-mediated processes that are only capable of transferring a substance from a region of higher concentration to one of lower concentration. These processes are sometimes also referred to as equilibrating carrier systems inasmuch as net transport ceases when the concentrations of the transported solute are the same on the two sides of the membrane, that is, when the system is equilibrated with respect to the solute in question. Thus, facilitated diffusion resembles noncarrier-mediated diffusional processes in that the direction of net flow is always downhill. It differs from them in that it exhibits all of the characteristics of carrier-mediated processes and often results in the transmembrane transfer of a solute that could not otherwise permeate the membrane; indeed, the latter is its sole function.

The classic example of facilitated diffusion is glucose transport across the membranes surrounding many animal cells, such as erythrocytes, striated muscle, and adipocytes. The glucose concentrations in these cells are much lower than the glucose concentration in the extracellular fluid because glucose is rapidly metabolized by these cells after gaining entry. The only well-documented exceptions to this statement are renal proximal tubular cells, small intestinal epithelial cells, and the cells of the choroid plexus. These cells are responsible for transepithelial glucose absorption, and their intracellular glucose concentrations may exceed those in the extracellular fluid; as is discussed below, mechanisms other than facilitated diffusion are responsible for the uptake of hexoses by these cells. Thus, for most cells the problem is not that of transporting glucose against a concentration difference but of transporting glucose rapidly across an essentially impermeant barrier. This is accomplished by a carrier mechanism, schematically illustrated in Fig. 6.2. Glucose (symbolized by the small circle labeled *S*) combines with the carrier from one side of the membrane to form a glucose–carrier complex. This is followed by a change in conformation of the complex that permits glucose to dissociate from the carrier and enter the solution on the other side of the membrane. The free carrier site is then available for another passenger. Since there are a limited number of carriers, the process is saturable and subject to competitive inhibition.

One of the important features of the carrier model for facilitated diffusion that is illustrated in Fig. 6.2 is that the carrier itself is unaltered during the translocation process, and only thermal energy is required for the conformational change that exposes the binding site to one or the other side of the membrane. Thus, the transport process is symmetrical, and it is just as easy for the solute *S* to move from the extracellular fluid into the cell as in the opposite direction. Consequently, when the concentrations of solute on the two sides of the membrane are equal, the system is

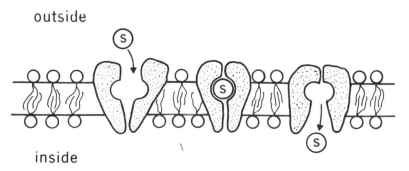

FIG. 6.2. Model for carrier-mediated facilitated diffusion of a solute *S*.

TABLE 6.1. *Facilitated transporters of glucose*

Name	Distinguishing characteristics
GluT 1	Widespread distribution but particularly high levels in plasma membranes of endothelial cells lining blood vessels and the blood–brain barrier
GluT 2	Basolateral membranes of small intestinal and renal proximal tubule epithelial cells
GluT 3	Has very high affinity for glucose and is present in plasma membranes of neurones
GluT 4	Present in plasma membranes of muscle cells and adipocytes and its expression is upregulated by insulin
GluT 5	Small intestinal and renal epithelial cells

entirely symmetrical, and the carrier-mediated flows in both directions will be equal. This is the reason why this transport process is not capable of bringing about net transport from a region of lower concentration to one of higher concentration and why net transport ceases when the solute distribution is equilibrated.

To date, five carrier proteins capable of mediating the facilitated diffusion of glucose have been identified. All five consist of a polypeptide chain composed of approximately 500 amino acids and possess a high degree of homology including twelve putative transmembrane spanning segments. Some of the distinguishing characteristics of these glucose transport proteins, which are abbreviated GluT, are given in Table 6.1.

Active Transport

Active transport is the term reserved for carrier-mediated transport processes that are capable of bringing about the net transfer of an uncharged solute from a region of lower concentration to one of higher concentration *or* the transfer of a charged solute against combined chemical and electrical driving forces.

Thus, active transport processes are capable of counteracting or reversing the direction of diffusion, a spontaneous process, and therefore are capable of performing work. The concept that active transport processes perform work may be difficult to grasp for those who have not had some acquaintance with thermodynamics. Because this is an extremely important concept, we digress for a moment and attempt to provide it an intuitive basis.

The direction of all natural change in the universe is for systems to move from a state of higher energy to one of lower energy. Thus, an unsupported weight will fall from a position of higher gravitational (or potential) energy to one of lower gravitational energy; electrons will flow through a conductor from a region of electronegativity (the cathode of a battery) to one that is electropositive (the anode); uncharged solutes will diffuse from a region of higher concentration to one of lower concentration. All of these processes are spontaneous inasmuch as they are accompanied by a decrease in the free energy of the system and do not require any external assistance or intervention; they will occur in a completely isolated system. It is a universal experience (and one of the basic tenets of thermodynamics) that once a spontaneous change has taken place, the initial conditions cannot be restored without an investment of energy; that is, the only way one can reverse a spontaneous process is by performing work. In the examples cited above, mechanical work is required to restore the weight to its original height, and electrical work is needed to recharge the battery. A thoroughly mixed solution can be "unmixed" by ultracentrifugation, ultrafiltration through an appropriate molecular sieve, or distillation; but, whatever means are chosen, it is clear that "unmixing" will never occur spontaneously and that the result of diffusion can only be reversed through the investment of energy.

The means by which a biological cell "reverses" diffusional flows is termed *active transport*, and here, too, work performed, and energy derived from metabolism, must be invested.

It follows that the ability of a cell to carry out active transport processes is dependent on an intact supply of metabolic energy, and all active transport processes can be inhibited by deprivation of essential substrates or through the use of metabolic poisons. Indeed, the *sine qua non* of active transport is the presence of a direct or indirect linkup or *coupling* between the carrier mechanism and cell metabolism; when an active transport process is initiated or accelerated, there is a concomitant increase in the metabolic rate (as measured by glucose utilization, oxygen consumption, etc.), and inhibition of active transport results in a decrease in the metabolic rate.

There are two classes of active transport processes, termed *primary active transport* and *secondary active transport*, that differ in the ways they derive (or are coupled to) a supply of energy.

Primary Active Transport

Primary active transport implies that the carrier mechanism responsible for the movement of a solute against a concentration difference or a combined concentration and electrial potential difference (for the case of ions) is directly coupled to metabolic energy.

The best-studied primary active transport processes in animal cells are

1. The carrier mechanism found in virtually all cells from higher animals that is responsible for maintaining their low cell Na^+ and high cell K^+ concentrations.
2. The carrier mechanisms found in sarcoplasmic reticulum and many plasma membranes responsible for active transport of Ca^{2+}.
3. The carrier mechanisms capable of actively extruding protons from the cells present in the gastric mucosa and renal tubule and actively pumping protons into intracellular organelles (e.g., lysozomes).

These carrier proteins have been purified, and all possess adenosine triphosphatase (ATPase) activity. That is, the same protein that is involved in the binding and translocation of Na^+ and K^+ is also capable of hydrolyzing adenosine triphosphate

(ATP) and somehow utilizing the chemical energy released to perform the work of transport. The same holds for the Ca^{2+} pumps and the proton pumps. The precise mechanisms whereby the chemical energy of the terminal phosphate bond of ATP is converted into transport work is not clear.

Secondary Active Transport

Secondary active transport refers to processes that mediate the uphill movements of solutes but are not directly coupled to metabolic energy; instead, the energy required is derived from coupling to the downhill movement of another solute. Let us illustrate such systems by considering Fig. 6.3.

Figure 6.3A portrays a rotating carrier molecule C within a membrane that has two binding sites, one for Na^+ and the other for a solute S, which, at this instant, are shown facing compartment o. Let us assume that the carrier can rotate only when both binding sites are empty or filled and is immobile when only one site is filled. Thus, it can transport Na^+ and S from compartment o to compartment i in a one-to-one fashion. Now let us assume that there is no (or very little) Na^+ in compartment i and that every Na^+ that enters from compartment o is "removed."

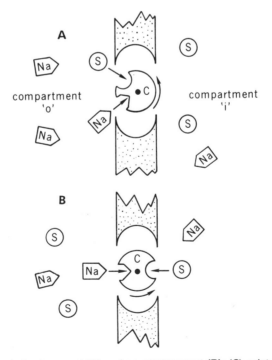

FIG. 6.3. Na^+-coupled cotransport **(A)** and countertransport **(B)**; (S) solute; (C) carrier molecule.

Clearly, under these conditions, S can only move (to any appreciable extent) from compartment o to compartment i; because there is no Na^+ in compartment i, S that enters this compartment cannot move back out and is "trapped." In time, the concentration of S in compartment i will exceed that in compartment o so that the system will have actively transported S without a direct linkup to metabolic energy. This system is referred to as *secondary active cotransport*.

How is this possible? We have stipulated that every Na^+ that enters compartment i is removed. In animal cells, this is accomplished by the (Na^+-K^+) pump that is directly linked to (energized by) ATP hydrolysis. Thus, in essence, energy is directly invested into a primary active transport mechanism that is responsible for extruding Na^+ from the cell (in exchange for K^+) and thereby maintaining a low intracellular Na^+ concentration. The $Na-S$ cotransport mechanism can then bring about the uphill movement of S energized by the downhill flow of Na^+. There are many Na^+-coupled secondary active cotransport processes in animal cell membranes. They include sugar and amino acid uptake across the apical membranes of small intestinal and renal proximal tubule cells; the uptake of many L-amino acids by virtually all non-epithelial cells; and Cl^- uptake by a variety of epithelial and nonepithelial cells.

Figure 6.3B illustrates another mechanism for secondary active transport that operates along similar principles. Let us assume that the carrier C has two sites, one

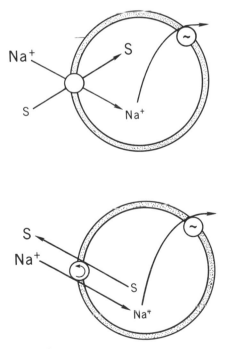

FIG. 6.4. Cellular models of Na^+-coupled cotransport (*top*) and countertransport (*bottom*); (*S*) solute.

facing compartment o and the other compartment i. Let us further assume that the carrier can rotate only when both sites are either empty or filled but not when only one site is filled. Clearly, this mechanism can bring about a one-to-one exchange of Na^+ for S across the membrane. If there is little or no Na^+ in compartment i, the system will only be able to exchange Na^+ in compartment o for S in compartment i and not vice versa. Thus, the downhill movement of Na^+ from o to i can bring about uphill flow of S from i to o; this mechanism is referred to as *secondary active countertransport*. And, once more, the trick is the removal of Na^+ from compartment i, which is accomplished by the primary active Na^+ transport mechanism that is energized by ATP hydrolysis. Two examples of such a countertransport system are (1) $Na^+–H^+$ exchange and (2) $Na^+–Ca^{2+}$ exchange; both of these mechanisms have been found in a wide variety of cell types. In both instances, H^+ and Ca^{2+} are extruded from the cell coupled to the downhill influx of Na^+ into the cell.

Cellular models of these co- and countertransport processes are illustrated in Fig. 6.4. The essential common feature of these transport processes is that metabolic energy is directly invested into the operation of the $(Na^+–K^+)$ pump, a primary active transport mechanism. The operation of this pump results in a cell Na^+ concentration that is much lower than that in the extracellular fluid. This transmembrane Na^+ gradient in turn provides the energy for many Na^+ coupled secondary active co- and countertransport processes.[1]

BIBLIOGRAPHY

Leinhard G E et al. How cells absorb glucose. *Sci Amer* 1992:86–91.
Reuss L, Russell JM, Jennings ML. *Molecular biology and function of carrier proteins.* New York; The Rockefeller University Press, 1992.
Schultz SG, Curran PF. Coupled transport of sodium and organic solutes. *Physiol Rev* 1970;50:637–718.
Semenza G, Kinne R. Membrane transport driven by ion gradients. *Ann N Y Acad Sci* 1985:456.
Stein WD. *Transport and diffusion across cell membranes.* Orlando: Academic Press, 1986; Chapters 4, 5, and 6.

---

[1] The terms *symport* and *antiport* are sometimes employed to describe co- and countertransport processes, respectively.

7

The "Pump–Leak" Model, the Origin of Transmembrane Electrical Potential Differences, and the Maintenance of Cell Volume

All biological membranes possess an assortment of pathways that permit the diffusion of water-soluble solutes, mainly ions (i.e., "leak pathways"), and carriers. These components, acting in concert, are responsible for the uptake of essential nutrients and building blocks by cells, the extrusion of the end products of some metabolic processes from cells, the maintenance of a near-constant (time-independent, or steady-state) intracellular composition and volume, and the establishment of transmembrane electrical potential differences.

In this chapter, we illustrate the interactions among carrier-mediated "pumps" and channel-mediated "leaks" by considering the processes responsible for the maintenance of the high intracellular concentrations of K^+ and the low intracellular concentrations of Na^+ characteristic of virtually all cells in higher animals. We have chosen this system as a prototype for other "pump-leak" systems, not only because of its ubiquity but also because of the essential role it plays in energizing a wide variety of other secondary active pumps, in essential bioelectric processes, and in the maintenance of cell volume.

THE (Na^+–K^+) PUMP

The fact that cells from higher animals contain a high K^+ concentration and a low Na^+ concentration compared to the extracellular fluid was established in the twentieth century shortly after analytic techniques for measuring these elements were developed. In the period 1940 to 1952, Steinbach demonstrated that when frog striated muscle is incubated in a K^+-free solution, the cells simultaneously lose K^+

and gain Na$^+$, and that this could be reversed by the addition of K$^+$ to the extracellular fluid. In the ensuing decade, abundant evidence accrued for the presence of carrier-mediated processes in biological membranes that bring about the extrusion of Na$^+$ from cells obligatorily coupled to the uptake of K$^+$ by cells that are directly dependent on adenosine triphosphate (ATP); this mechanism is referred to as the (Na$^+$–K$^+$) pump. Furthermore, it appears that, in many cells, three Na$^+$ ions are extruded in exchange for two K$^+$ ions for each ATP consumed.

In 1958, Skou identified an ATPase in a homogenate of crab nerve tissue whose hydrolytic activity was dependent on the simultaneous presence of Na$^+$ and K$^+$ in the assay medium. In addition, ATPase activity in the presence of Na$^+$ and K$^+$ could be inhibited by glycosides derived from the wild flower *Digitalis purpurea* (foxglove) (e.g., ouabain), known since 1953 to be potent inhibitors of the carrier-mediated transport mechanisms responsible for the "active" extrusion of Na$^+$ from cells coupled to the "active" uptake of K$^+$ [i.e., the (Na$^+$–K$^+$) pump].

During the past four decades, innumerable studies have established that the (Na$^+$–K$^+$) pump and the (Na$^+$, K$^+$)ATPase are one and the same. Furthermore, this (pump)ATPase has been isolated, purified, and reconstituted in active form in artificial lipid vesicles. It is now clear that it consists of two subunits: α and β. The α subunit has a molecular weight of approximately 100,000 daltons, is minimally, if at all, glycosolated, and is the subunit that possesses the ATPase (catalytic) activity as well as the ability to bind Na$^+$, K$^+$, and digitalis glycosides such as ouabain. The β subunit has a molecular weight of about 55,000 daltons, of which approx-

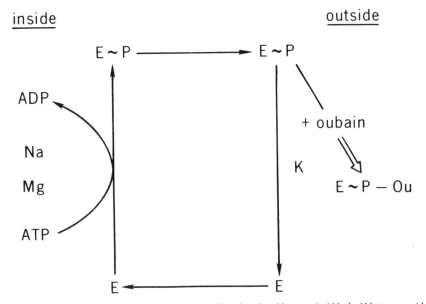

FIG. 7.1. Simplified schematic of the partial reactions involved in coupled Na$^+$–K$^+$ transport by (Na$^+$, K$^+$)ATPase. Ou, oubain.

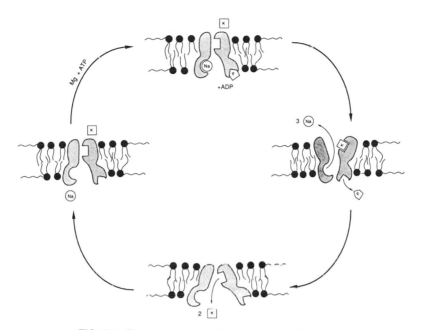

FIG. 7.2. The possible operation of the (Na^+–K^+) pump.

imately two thirds can be attributed to polypeptides and one third to glycosolation; the β subunit has no ATPase activity, and its function may be to direct and insert (or anchor) the α subunit to the plasma membrane. Coassociation of the α and β subunits is necessary for pump activity.

The results of studies on the biochemical behavior of this ATPase are consistent with the *simplified* sequence of events illustrated in Fig.7.1. In the presence of intracellular Na^+ and Mg^{2+}, the (Na^+, K^+)ATPase (E) is capable of hydrolyzing ATP to form a high-energy intermediate (E~P) that is capable of binding Na^+. The interaction between (E~P) and extracellular K^+ results in its hydrolysis, thereby reforming E and completing the cycle. Thus, the recycling of this enzyme-pump from the E stage through (several) high-energy (E~P) stages back to the original (E) stage requires the simultaneous presence of Na^+ in the intracellular compartment and K^+ in the extracellular compartment. Digitalis glycosides (such as ouabain) appear to bind tightly with (E~P) at or near the site where K^+ interacts with the enzyme and thereby prevent the conversion of (E~P) to E; this aborts the cycle. This sequence of events is shown in Fig. 7.2.

THE PUMP-LEAK MODEL

The plasma membranes surrounding cells of higher animals not only contain (Na^+–K^+) pumps but are also traversed by channels that permit the diffusional

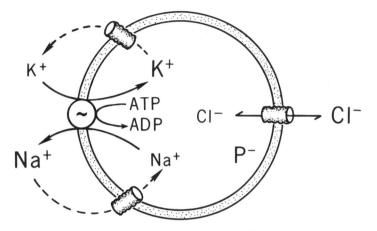

FIG. 7.3. Model of a cell containing an ATP-dependent (Na⁺–K⁺) exchange pump, leak pathways for Na⁺ and K⁺, and a pathway for Cl⁻ diffusion across the membrane; P⁻ denotes negatively charged intracellular macromolecules (mainly phosphates and proteins). Reasonable values for intracellular and extracellular K⁺ are 140 and 5 mM, respectively; for intracellular and extracellular Na⁺, 15 and 140 mM, respectively.

flows (leaks) of Na⁺ and K⁺ across those barriers. The interaction between these pumps and leaks is illustrated in Fig. 7.3. Briefly, the (Na⁺–K⁺) pumps extrude Na⁺ from the cells and simultaneously propel K⁺ into the cells at the expense of metabolic energy (ATP). This results in a low intracellular Na⁺ concentration and a high intracellular K⁺ concentration, which in turn sets the stage for Na⁺ diffusion into the cell and K⁺ diffusion out of the cell through their respective leak pathways. The final result of the interactions between the pump and leaks is a time-independent or steady-state condition, where the movements of Na⁺ and K⁺ mediated by the pumps are precisely balanced by the oppositely directed flows of these ions through their leak pathways.

This pump-leak system is present in all cells in higher animals and is responsible for maintaining the low cell Na⁺ and high cell K⁺ concentrations characteristic of those cells. As discussed previously, the resulting Na⁺ gradient (or concentration difference) across the plasma membrane can serve to energize a number of secondary active transport processes, such as the Na⁺-coupled accumulation of amino acids (cotransport), the extrusion of H⁺ produced by metabolic processes via the Na⁺–proton countertransport mechanism, and the regulation of cell Ca²⁺ by the Na⁺–Ca²⁺ countertransport mechanism. In the final analysis, all of these secondary active transport processes derive their energy from the ATP hydrolyzed by the (Na⁺–K⁺) pump—a truly remarkable design.

TRANSMEMBRANE ELECTRICAL POTENTIAL DIFFERENCES

All biological membranes are characterized by transmembrane electrical potential differences, which are, in almost all instances, oriented so that the cell interior is electrically negative with respect to the extracellular compartment. The size of the electrical (membrane) potential difference, V_m, ranges from -10 mV to as high as about -100 mV. Furthermore, in a number of cell types Vm is variable, and this variation is responsible for the propagation of signals by nerve tissue, contraction of muscle, and stimulus-secretion coupling in exocrine and endocrine secretory cells.

What Is the Origin of V_m?

To appreciate how transmembrane electrical potential differences arise as a consequence of the interaction between pumps and leaks, let us consider a hypothetical cell, such as that illustrated in Fig. 7.4, which contains a coupled potassium acetate (KAc) pump energized by the hydrolysis of ATP and leak pathways for K^+ and Ac^-. Let us assume that this cell is initially filled with distilled water and is then dropped into a solution with a concentration $[KAc]_o = 10$mM. Initially, KAc will enter the cell, some via the pump and some via diffusion through the leaks. When the intracellular concentration reaches 10mM, diffusion of K^+ and Ac^- into the cell will cease. But, because the pump continues to operate, in time the concentration of KAc in the cell will exceed that in the extracellular solution, and then K^+ and Ac^- will diffuse out of the cell through their leak pathways. At a subsequent time, a point will be reached when the rate at which KAc is pumped into this cell is precisely balanced by the rates at which K^+ and Ac^- diffuse out of the cell, and let us say for the sake of discussion that when this steady state is reached, the concentration of KAc in the cell $[KAc]_i = 100$ mM. Furthermore, since the stoichiometry of the pump is one to one, the rates at which K^+ and Ac^- diffuse out of the cell must also be equal or the law of electroneutrality will be violated.

Since the concentration differences for K^+ and Ac^- across the membrane are equal (i.e., $\Delta C_K = \Delta C_{Ac} = 90$ mM), if the membrane is equally permeable to K^+ and Ac^- (i.e., $P_K = P_{Ac}$), equal rates of diffusion out of the cell (i.e., $J_K = J_{Ac}$)

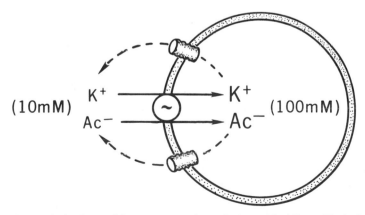

FIG. 7.4. Hypothetical cell containing an energy-dependent pump for KAc and leaks for K^+ and Ac^-.

would not pose a problem. But, what if the membrane is more permeable to the small ion, K^+, than to Ac^-? Then, K^+ will tend to leave the cell faster than Ac^- and possibly cause a violation of the law of electroneutrality. This is prevented by the establishment of an electrical potential difference across the membrane oriented such that the cell interior is electrically negative with respect to the extracellular solution. This retards the diffusion of K^+ out of the cell and accelerates the outward diffusion of Ac^- so that $J_K = J_{Ac}$.

This situation is precisely analogous to that which we considered in Chapter 4 dealing with the diffusion of KAc across an artificial membrane separating two solutions having different concentrations. The only differences are as follows:

1. In the artificial system, we provided the "muscle" (energy) to set up the concentration difference, whereas in our hypothetical cell, the energy is provided by the pump at the expense of ATP.
2. The artificial system will "run down" in time as KAc diffuses out of compartment o into compartment i, whereas in our hypothetical cell, the KAc leaving the cell is constantly being replenished by the pump; this is the essence of a steady state displaced from equilibrium by the investment of metabolic energy.

Finally, we can estimate V_m by employing Eq. 4.2, which describes the diffusion potential across a membrane arising from a concentration difference of a dissociable salt and differences in the permeabilities of the membrane to the resulting ions; that is,

$$V_m = \left\{ \frac{P_K - P_{Ac}}{P_K + P_{Ac}} \right\} \times 60 \log \left\{ \frac{[KAc]_o}{[KAc]_i} \right\} \text{ mV} \qquad [7.1]$$

(Note: In Eq. 4.2 we employed diffusion constants, but the use of permeability coefficients as defined in Eq. 2.5 is acceptable.)

Thus, if $P_K = P_{Ac}$, then $V_m = 0$. But if $P_K > P_{Ac}$, then, because $[KAc]_i > [KAc]_o$, V_m will be oriented such that the cell interior is electrically negative with respect to the extracellular solution. Furthermore, since $([KAc]_o/[KAc]_i) = 0.1$ (i.e., 10 mM/100 mM), if P_K is very much greater than P_{Ac}, then V_m will approach -60 mV, that is, the Nernst potential for K^+. If $P_{Ac} > P_K$, then the cell interior will be electrically positive with respect to the extracellular solution; in this case, V_m can approach $+60$ mV if P_{Ac} is very much larger than P_K. Thus, V_m can assume any value between approximately $+60$ mV and approximately -60 mV, depending on the relation between P_K and P_{Ac}. Recall that in no instance is bulk electroneutrality violated; indeed, it is V_m, which assures that $J_K = J_{Ac}$, that prevents a bulk separation of charge.

Now let us examine the behavior of the somewhat more realistic model illustrated in Fig. 7.3. Suppose that the only two ions subject to active transport are Na^+ and K^+ and that a one-to-one coupled carrier mechanism is involved; Cl^- is assumed to cross the cell membrane only by diffusion through water-filled pores. Under steady-state conditions, (1) Na^+ must diffuse into the cell at a rate equal to the rate of its carrier-mediated extrusion, and (2) K^+ must diffuse out of the cell at a rate equal to the rate of its carrier-mediated uptake. If the Na^+ and K^+ carrier mechanism is coupled such that for every K^+ pumped into the cell one Na^+ is extruded, the diffusional flows of these ions in opposite directions must also be equal, otherwise electroneutrality would be violated. Now in the example given, the concentration differences for Na^+ and K^+ across the membrane are approximately equal and opposite (this is approximately the case for many cells in higher animals) so that, if the permeabilities of the membrane to Na^+ and K^+ were also equal, the rates of diffusion in opposite directions would be equal and electroneutrality would be preserved. However, if the membrane is much more permeable to K^+ than to Na^+ (as is the case for most cells), the rates of net Na^+ and K^+ diffusion would not be equal in the absence of a transmembrane electrical potential difference. Thus, a V_m is generated that it is oriented such as to retard the diffusion of the more permeant ion (K^+) and accelerate the diffusion of the less permeant ion (Na^+) so that the inward diffusion of Na^+ is equal to the outward diffusion of K^+. Under these conditions, the cell interior will be electrically negative with respect to the exterior. To repeat, a diffusion potential arises across the membrane to maintain equal diffusion rates and to preserve electroneutrality despite the fact that the permeabilities of the membrane to the two ions are not equal.

Several important points should be noted:

1. The example dealing with Na^+ and K^+ diffusion in opposite directions is formally analogous to an earlier example (Fig. 7.4) in which K^+ and Ac^- diffused in the same direction, because, as far as electroneutrality is concerned, the flow of a cation in one direction is equivalent to the flow of an anion in the opposite direction. In both examples, when K^+ is the more permeable ion, the orientation of V_m will be the same (i.e., so as to retard the flow of K^+ and accelerate the flow of the less permeable ion).

2. The electrical potential differences (V_m) described in both examples are diffusion potentials resulting from the diffusional flows of ions down concentration differences. Although V_m is dependent on ion pumps, the relation is indirect; the pumps merely serve to establish the ionic concentration differences that provide the driving forces for the diffusional flows.

3. The orientation of V_m is always such as to retard the diffusional flow of the more permeant ion and accelerate the flow of the less permeant ion, and the magnitude of V_m is dependent on the individual permeabilities and concentrations of these ions, because these are the direct determinants of the diffusional flows that must be equalized by this electrical potential difference. We can generate an expression that defines the magnitude and orientation of V_m using an argument similar to that employed when we considered the diffusion potential generated by KAc diffusion out of the hypothetical cell shown in Fig. 7.4. Thus, if P_K is much greater than P_{Na} then V_m will approach the Nernst potential for K^+ {i.e., $60 \log ([K^+]_o/[K^+]_i)$}; and, because $[K^+]_o < [K^+]_i$ the cell interior will be electrically negative with respect to the extracellular compartment (i.e., $V_m < 0$). Conversely, if P_K is much less than P_{Na}, then V_m will approach the Nernst potential for Na^+ {i.e., $60 \log ([Na^+]_o/$

$[Na^+]_i)\}$; and, because $[Na^+]_o > [Na^+]_i$, under these conditions $V_m > 0$. These two extremes are satisfied by Eq. 7.2, which is sometimes referred to as a "double Nernst equation":

$$V_m = 60 \log \left\{ \frac{P_K[K^+]_o + P_{Na}[Na^+]_o}{P_K[K^+]_i + P_{Na}[Na^+]_i} \right\} \qquad [7.2]$$

This equation can be rewritten in the form:

$$V_m = 60 \log \left\{ \frac{[K^+]_o + \alpha[Na^+]_o}{[K^+]_i + \alpha[Na^+]_i} \right\} \qquad [7.3]$$

where $\alpha = (P_{Na}/P_K)$. When P_K is much greater than P_{Na}, then α is very small and V_m approaches the Nernst potential for K^+; conversely, when P_K is much less than P_{Na} then α is very large and V_m approaches the Nernst potential for Na^+. Thus, V_m can have any value between these two extremes depending on the value of α.

An expression having the form of Eq. 7.2 was first formally derived by Goldman in 1943. It was rederived by Hodgkin and Katz in 1949 and, as will be discussed in Chapter 8, was first applied successfully to the analysis of the ionic basis of the resting and action potential of the squid axon. During the past three decades, it has provided the basis for understanding the origin of V_m across a wide variety of biological membranes.

4. Finally, it should be noted that the stoichiometry of the (Na–K) pump is three Na^+ for two K^+ (not one to one). All this means is that, when a steady state is achieved, three Na^+ ions must diffuse into the cell for every two K^+ ions that diffuse out of the cell. But, if $P_{Na} \ll P_K$, the final result will still be that V_m is oriented such that the cell interior is electrically negative with respect to the extracellular fluid.

Before concluding this section, let us consider the distribution of Cl^- resulting from this pump-leak system. If, as stipulated above, the Cl^- distribution is determined solely by diffusion, then, when there is no net movement of Cl^- across the membrane (i.e., the composition of the cell is constant), the chemical potential forces acting on Cl^- must be balanced by the electrical forces acting on this anion. Thus, if V_m is negative (with respect to the outer solution), then the intracellular concentration of Cl^- (i.e., C_{Cl}^i) will be less than that in the extracellular solution (i.e., C_{Cl}^o). The precise relation among V_m, C_{Cl}^i, and C_{Cl}^o is given by the Nernst equation, which, as discussed above, describes the condition for the balance or equality of chemical and electrical forces:

$$V_m = 60 \log[C_{Cl}^i/C_{Cl}^o] \qquad [7.4]$$

or, transposing,

$$[C_{Cl}^i/C_{Cl}^o] = 10^{[V_m/60]} \qquad [7.5]$$

When V_m is negative, the right-hand side of Eq. 7.5 has a value less than 1 so that $C_{Cl}^i < C_{Cl}^o$. Thus, if V_m is approximately -60 mV, then C_{Cl}^i will be approximately one-tenth C_{Cl}^o.

ELECTROGENIC (OR RHEOGENIC) ION PUMPS

It is important to stress once more that the transmembrane potential differences we have discussed to this point do not arise directly from the (Na^+-K^+) pump but indirectly from the ionic asymmetries that are generated by this pump. It is now clearly established, however, that under most circumstances, the (Na^+-K^+) pump is not neutral (i.e., one to one) but actually extrudes three Na^+ from the cell in exchange for two K^+. Thus, the pump itself brings about the movement of charge across the membrane (i.e., one positive charge from inside to outside for every cycle) and can be viewed as a current generator. Clearly, inasmuch as the pump generates a current across a membrane with a finite resistance, its action must directly result in an electrical potential difference given by the product of the "pump current" and the membrane resistance.

Thus, in general, transmembrane electrical potential differences have two origins. By far the largest fraction of these potential differences is attributable to diffusion potentials due to ionic asymmetries established by (Na^+-K^+) pumps. But, definite contributions to these potential differences arise directly from the current generated by the (Na^+-K^+) pump.

TRANSMEMBRANE DISTRIBUTION OF SOLUTES UNDER STEADY-STATE CONDITIONS

The above considerations provide us with the principles that permit us to deduce important information with regard to the nature of the distributions of metabolically inert solutes across cell membranes when the cell composition is constant (i.e., in a steady state).

Thus, let us assume that we can measure the electrical potential difference across a cell membrane, V_m, by employing the microelectrophysiological techniques described in Chapter 8 and at the same time determine the intracellular and extracellular concentrations (or, more properly, activities) of any solute i (i.e., C_i^i and C_i^o, respectively).

Now, the Nernst equation provides us with the criterion for determining whether the ratio of the concentrations (activities) of i across the membrane can be attributed entirely to thermal ("passive") forces or whether additional ("active") forces are necessary. Thus, if (at 37°C),

$$V_m = E_i = (60/z_i) \log(C_i^o/C_i^i) \ (mV) \qquad [7.6]$$

or, multiplying both sides of Eq. 7.6 by z_i,

$$z_i V_m = 60 \log(C_i^o/C_i^i) \ (mV) \qquad [7.7]$$

then, the steady-state distribution of i across the membrane can be considered passive. If this equality does not hold, then forces in addition to thermal energy must be involved in establishing the observed *distribution ratio* of i.

Now, we can rearrange Eq. 7.7 to provide the somewhat more useful expressions

$$(C_i^o/C_i^i) = 10^{(z_iV_m/60)} \quad \text{or} \quad (C_i^i) = (C_i^o)\,10^{(-z_iV_m/60)} \qquad [7.8]$$

Thus, if we know V_m and the extracellular concentration of solute i, C_i^o, we can predict the intracellular concentration C_i^i that would be consistent with a passive distribution of i across the membrane and then compare that predicted value with the actual measured value.

For a neutral solute (i.e., $z_i = 0$), Eq. 7.8 states that the distribution of i across the cell membrane is independent of V_m and that if $C_i^o = C_i^i$, this distribution can be accounted for by passive transport processes that do not require direct or indirect coupling to a source of metabolic energy. If $C_i^i > C_i^o$, then metabolic energy must be invested into the transport process to "pump" i into the cell. Conversely, if $C_i^i < C_i^o$, then energy must be invested to extrude i from the cell. Thus, for a neutral, inert solute, if $C_i^i \neq C_i^o$, transport across the cell membrane cannot be attributed to diffusion or carrier-mediated, facilitated diffusion.

Now let us turn to charged solutes and illustrate this approach by considering a cell bathed by a plasma-like solution, where $C_{Na}^o = 140$ mM, $C_K^o = 4$ mM, $C_{Cl}^o = 120$ mM, and $C_{Ca}^o = 2$ mM. The electrical potential difference across this membrane is determined to be -60 mV, and the intracellular concentrations of Na^+, K^+, Cl^-, and Ca^{2+} are determined to be 10, 120, 12, and 10^{-3} mM, respectively.

With respect to Na^+, Eq. 7.8 predicts that the intracellular concentration consistent with a passive distribution should be $C_{Na}^i = 10C_{Na}^o = 1{,}400$ mM; but the observed value of C_{Na}^i was only 10 mM. Thus the distribution of Na^+ across the cell membrane cannot be attributed to passive transport processes. Instead, energy must be invested by the cell to extrude Na^+ and thereby lower its intracellular concentration to a level well below that predicted for a simple passive distribution.

Turning to K^+, Eq. 7.8 predicts that if K^+ is passively distributed across the membrane, its intracellular concentration should be $C_K^i = 10C_K^o = 40$ mM. The observed intracellular concentration (120 mM) is much greater than this predicted value so that energy must be invested to actively pump K^+ into the cell.

With respect to Cl^-, the predicted value for C_{Cl}^i is $0.1C_{Cl}^o$, or 12 mM. This agrees with the actually measured value so that one can conclude that the distribution of Cl^- across the cell membrane is the result of passive transport processes that do not require an investment of energy on the part of the cell.

Finally, applying Eq. 7.8 to the case of Ca^{2+}, we see that its predicted intracellular concentration is $C_{Ca}^i = 2 \times 10^2 = 200$ mM. This predicted value is much greater than the observed value of only 10^{-3} mM so that the cell must invest energy into active transport processes to extrude Ca^{2+} from its interior. In short, the application of the Nernst equation, which defines the thermal balance of chemical and electrical forces across a membrane, permits us to determine whether the distribution of any inert solute across a cell membrane is passive or active.

Finally, there are two caveats with regard to the application of the Nernst equation to the steady-state distributions of solutes across cell membranes. First, non-conformity with the predictions of the Nernst equation simply means that the observed distribution of a solute cannot be attributed to passive forces alone, but it provides no insight into the detailed mechanism(s) responsible for this active distribution.

Second, this line of reasoning does not apply to solutes that are either produced or utilized by cells. For example, the steady-state glucose concentration in most cells is much lower than that in the extracellular fluid because this nutrient is rapidly metabolized after gaining entry into the cells. As another example, the steady-state concentration of urea in most cells is greater than that in the extracellular fluid inasmuch as it is produced by protein catabolism and exits the cell by passive transport processes; thus, under steady-state conditions, there must be a concentration difference across the membrane that provides the driving force for urea exit at a rate equal to that at which it is produced. Clearly, an uncritical application of the criteria discussed above to these two neutral solutes would suggest that glucose is actively extruded from the cell and that urea is actively accumulated by these cells.

THE GIBBS-DONNAN EQUILIBRIUM, ION PUMPS, AND MAINTENANCE OF CELL VOLUME

So far we have considered three of the functions that are fulfilled by the operation of (Na^+-K^+) pumps. First, they are responsible for the high K^+ and low Na^+ concentrations characteristic of the intracellular fluid of higher animals. A number of enzymes involved in intermediary metabolism and protein synthesis appear to require relatively high concentrations of K^+ for optimal activity and are inhibited by high concentrations of Na^+; the activities of these enzymes would be markedly impaired if the intracellular Na^+ and K^+ concentrations were the same as those in the extracellular fluid. Second, these ionic asymmetries are largely responsible for establishing the electrical potential differences across membranes and the bioelectrical phenomena essential for the functions of nerve and muscles. Finally, the Na^+ gradients established by these pumps energize the secondary active transport of a number of solutes whose movements are coupled to those of Na^+ (cotransport or countertransport).

(Na^+-K^+) pumps fulfill another, extremely vital, function in cells that do not possess a rigid cell wall; namely, they are in part responsible for maintaining the intracellular osmolarity equal to that of the extracellular fluid, thereby preventing osmotic water flow into the cells and, in turn, cell swelling.

To appreciate why ion pumps are necessary for the maintenance of cell volume, we should first consider the artificial system illustrated in Fig. 7.5. The two compartments illustrated in Fig. 7.5 are assumed to be closed to the atmosphere and separated by a membrane that is freely permeable to Na^+, Cl^-, and water but is impermeable to proteins. A solution of NaCl is added to compartment o, and a solution of the sodium salt of a protein [Na^+-proteinate (NaP)], is added to compartment i; assume that at the outset $C_{Na}^i = C_{Na}^o$. The system is then left undisturbed for a sufficiently long time until equilibrium is achieved.

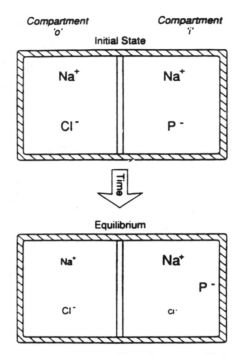

FIG. 7.5. The development of the Gibbs-Donnan equilibrium condition.

Let us now consider the characteristics of this final, time-independent, equilibrium condition, which was derived by Josiah Willard Gibbs (1839–1903) and Frederick George Donnan (1870–1956) and is often referred to as the *Gibbs-Donnan equilibrium*, the *Donnan equilibrium*, or the *Donnan distribution*.

Because the membrane is permeable to Na^+ and Cl^-, but there is no Cl^- initially present in compartment i, some Cl^- must diffuse from compartment o into compartment i; but this must be also accompanied by an equal amount of Na^+, otherwise the law of electroneutrality would be violated.

At all times, the system must obey a balance of electrical charges such that

$$C_{Na}^o = C_{Cl}^o \quad \text{and} \quad C_{Na}^i = C_{Cl}^i + z_p C_P^i \qquad [7.9]$$

where z_p is the valence of the protein molecule.

When equilibrium is achieved, the system will be characterized by three properties:

1. Because we started out with equal Na^+ concentrations in both compartments and Na^+ and Cl^- subsequently diffused into compartment i (at equal rates), the equilibrium condition must be characterized by asymmetrical distributions of both of these permeant ions across the membrane. As we have already learned, if there is an asymmetric distribution of a passively transported ion across a mem-

brane, then there must be an electrical potential difference, V_m, across that membrane that balances the concentration difference and is given by the Nernst equation. Thus, one value of V_m must simultaneously satisfy the equilibrium distributions of both Na^+ and Cl^-, namely,

$$V_m = 60 \log(C_{Na}^o/C_{Na}^i) = 60 \log(C_{Cl}^i/C_{Cl}^o) \quad [7.10]$$

2. It follows from Eq. 7.10 that

$$(C_{Na}^o/C_{Na}^i) = (C_{Cl}^i/C_{Cl}^o) \quad [7.11]$$

If we consider the initial and final (equilibrium) conditions illustrated in Fig. 7.5 together with Eq. 7.9, it should be clear that when equilibrium is achieved $C_{Na}^i > C_{Na}^o$ and $C_{Cl}^i < C_{Cl}^o$ with the precise relation given by Eq. 7.11. Furthermore, it follows from Eq. 7.10 that compartment i will be electrically negative with respect to compartment o.

3. Finally, if we consider the initial and final conditions together with Eq. 7.9, it should be clear that when equilibrium is achieved the concentration of osmotically active solutes in compartment i will be greater than that in compartment o, so that a pressure will have developed across the membrane given by the van't Hoff Eq. 3.1. Furthermore, if instead of being rigid the membrane is distensible, it would bulge into compartment o.

We chose the initial condition $C_{Na}^o = C_{Na}^i$ simply to make it easier for students to comprehend the evolution of the asymmetries that characterize this equilibrium condition. But these equilibrium characteristics can be formally generalized to any set of initial conditions. Thus, if compartment i has a greater concentration of impermeant charged species than compartment o, the three additional asymmetries that will characterize the Gibbs-Donnan equilibrium when it is reached are as follows:

1. There will be an asymmetric distribution of all permeant monovalent cations (C_+) and anions (C_-), which conforms to the relation

$$C_+^o/C_+^i = C_-^i/C_-^o = r$$

where r is often referred to as the *Donnan ratio* and is, in part, a function of the difference in total charge between compartments o and i borne by impermeant ions. If compartment i contains a preponderance of impermeant anions, then $C_+^i > C_+^o$ and $C_-^i < C_-^o$; if the impermeant species are predominantly cationic, then these relations will be reversed.

2. There will be an electrical potential difference across the membrane given by the relation

$$V_m = 60 \log(C_+^o/C_+^i) = 60 \log(C_-^i/C_-^o) = 60 \log r$$

If compartment i contains a preponderance of impermeant anions then V_m will be oriented such that compartment i is electrically negative with respect to compartment o; if the impermeant species are predominantly cationic, then this orientation will be reversed.

3. Regardless of the sign of the total charge carried by the preponderance of impermeant species in compartment i, at equilibrium that compartment will contain a greater number of osmotically active particles than compartment o. Thus, there will be an osmotic driving force for the movement of water into compartment i. If the membrane is rigid, then an osmotic pressure would balance that driving force; if the membrane is distensible, then it would bulge into compartment o.

We can appreciate the function of ion pumps in the maintenance of cell volume by considering what would happen if there were no ion pumps. Because the intracellular concentration of charged, largely anionic, impermeant macromolecules is much greater than that in the extracellular fluid, cells lacking ion pumps would resemble the passive system illustrated in Fig. 7.5 and would move toward the direction of achieving a Gibbs-Donnan equilibrium. The total osmotic activity of intracellular solutes would exceed that in the surrounding fluid, and there would be a driving force for osmotic water flow into the cell. If the cells possess rigid cell walls that prevent any increase in cell volume, an osmotic pressure difference would develop across the cell walls and the cell interiors would be subjected to a pressure greater than the extracellular fluid ("turgor"). If, as in the case of animal cells, the membrane is distensible, water would flow into the cell leading to cell swelling and, perhaps, rupture.

The (Na^+–K^+) pumps in animal cell membranes serve to reduce the intracellular content of osmotically active solutes, thereby counteracting the osmotic effect of intracellular macromolecules. The pumps extrude three sodium ions in exchange for two potassium ions and also establish an electrical potential difference across the membrane (negative cell interior) that reduces the steady-state intracellular concentrations of permeant, passively distributed anions (mainly Cl^-).

If the (Na^+–K^+) pumps are inhibited by digitalis glycosides or metabolic poisons, the cells will lose K^+ and gain Na^+ and Cl^-; in many cells, three Na^+ plus one Cl^- will be gained for every two K^+ lost, so that the total amount of Na^+ and Cl^- gained by the cell exceeds the amount of K^+ lost, and the total intracellular solute concentration will increase. This will result in an osmotic uptake of water, cell swelling, and may lead to the destruction of the integrity of the cell membrane.[1]

In summary, the membranes that surround most animal cells are distensible and highly permeable to water. If these cells are immersed in a hypertonic fluid, water rapidly leaves the intracellular compartment and they shrink. If these cells are immersed in a hypotonic solution, water flows rapidly into them and they swell, possibly rupturing. In higher animals, the osmolarity of the extracellular fluid is carefully regulated by the kidneys in response to neurohormonal stimuli so that it normally remains within very narrow limits. However, the maintenance of isotonicity between the intracellular and extracellular fluids depends, in part, on the presence of

---

[1]Recall that inhibition of the pump would not only lead to the dissipation of the asymmetric distributions of Na^+ and K^+, but also the V_m (cell interior negative) arising from these asymmetries. According to Eq. 7.8, if V_m becomes less negative, then C_{Cl}^i will increase. Also note that the redistribution of ions following inhibition of the pump does not (indeed, cannot) violate the law of bulk electroneutrality.

ion, particularly (Na^+-K^+), pumps in the cell membranes. These pumps serve to lower the intracellular concentration of permeant solutes and thereby balance and offset the osmotic effects of impermeant intracellular macromolecules.[2]

BIBLIOGRAPHY

Hoffman JF. Active transport of Na^+ and K^+ by red blood cells. In: TE Andreoli, JR Hoffman, DD Fanestil, and SG Schultz, eds. *Physiology of membrane disorders.* New York: Plenum Press, 1985; Chapter 13.

Kaplan JH. Sodium ions and the sodium pump: transport and enzymatic activity. *Am J Physiol*, 1983: 245:G327.

Schultz SG. *Basic principles of membrane transport.* Cambridge: Cambridge University Press, 1980.

Strange K. Cellular and molecular physiology of cell volume regulation. Boca Raton, FL: CRC Press, 1993.

---

[2]There are other transport mechanisms that come into play when the preservation of cell volume is threatened by conditions that lead to swelling and, in some instances, shrinking. A discussion of these mechanisms is beyond the scope of this introductory text but can be found in the references.

8

Resting Potentials and Action Potentials in Excitable Cells

As discussed in Chapter 7, resting potentials are characteristic features of all cells in the body. But, nerve cells and other excitable cells, such as muscle cells, not only have resting potentials but are capable of altering these potentials for the purpose of communication, in the case of nerve cells, and for the purpose of initiating contraction, in the case of muscle cells. The material that follows introduces the ionic mechanisms that endow excitable membranes with this ability.

EXTRACELLULAR RECORDING OF THE NERVE ACTION POTENTIAL

The existence of "animal electricity" was known for well over 200 years, but the first direct experimental evidence for it was not provided until the development of electronic amplifiers and oscilloscopes. Figure 8.1 illustrates one of the earliest recordings that demonstrated the ability of nerve cells to alter their electrical activity for the purpose of coding and transmitting information. In this experiment, performed in 1934 by Hartline, extracellular recordings were made from the optic nerve of an invertebrate eye. Details of the techniques and interpretation of these extracellular recordings are described in Chapter 11, but for now it is sufficient to know that it is possible to place an electrode on the surface of a nerve axon and record electrical events that are associated with potential changes taking place across the axonal membrane. In the experiment illustrated in Fig. 8.1, light flashes of different intensities were delivered to the eye. With a very weak intensity light flash, there was no change in the baseline electrical activity. When the intensity of the light flash was increased, however, small spikelike transient events associated with the onset of the light were observed. Increasing the intensity of the light flash produced an increase in the rate of these spikelike events. These spikelike events are known as nerve action potentials, impulses, or, simply, spikes.

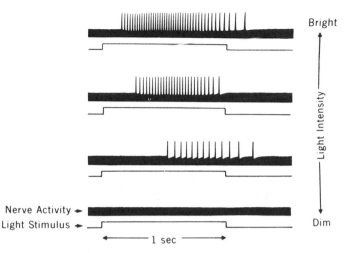

FIG. 8.1. Action potentials recorded from an invertebrate optic nerve in response to light flashes of different intensities. With dim illumination no action potentials are recorded, but with more intense illumination the number and frequency of action potentials increases. (Modified from H. K. Hartline, *J Cell Comp Physiol* 1934;5:229.)

Despite the fact that this experiment was performed more than 50 years ago, it nonetheless illustrates three basic properties of nerve action potentials and how they are utilized by the nervous system to encode information. First, nerve action potentials are very short, having a duration of only about 1 msec (1 msec = 10^{-3} sec). Second, action potentials are initiated in an *all-or-nothing* manner. Note that the amplitude of the action potentials does not vary during a sustained light flash. Third, and related to the above, with increasing stimulus intensity, it is not the size of action potentials that varies but rather their number or frequency. This is the general means by which intensity information is coded in the nervous system, and it is true for a variety of peripheral receptors. The greater the intensity of a physical stimulus (whether it be a stimulus to a photoreceptor, a stimulus to a mechanoreceptor in the skin, or a stimulus to a muscle receptor), the greater is the frequency of nerve action potentials. This finding has given rise to the notion of the frequency code for stimulus intensity in the nervous system.

Most of the information transmitted to the central nervous system from the periphery is mediated by nerve action potentials. Moreover, all the motor commands initiated in the central nervous system are propagated to the periphery by nerve action potentials, and action potentials produced in muscle cells are the first step in the initiation of muscular contraction. Action potentials are therefore quite important, not only for the functioning of the nervous system, but also for the functioning of muscle cells, and for this reason it is important to understand the ionic mechanisms that underlie the action potential and its propagation.

INTRACELLULAR RECORDING OF THE RESTING POTENTIAL

The action potentials illustrated on Fig. 8.1 were recorded with extracellular electrodes. To examine the properties of action potentials in greater detail, it was necessary to move from these rather crude extracellular techniques to intracellular recording techniques.

Figure 8.2 illustrates in a schematic way how it is possible to record the membrane potential of a living cell. The upper left of Fig. 8.2A is an idealized nerve cell, composed of a cell body with a portion of its attached axonal process. Outside the nerve cell in the extracellular medium is a glass microelectrode that is connected to a suitable voltage recording device, such as a voltmeter, a pen recorder, or an oscilloscope. A glass microelectrode is nothing more than a piece of thin capillary tubing that is stretched under heat to produce a very fine tip having a diameter less than 1 μm. The electrode is then filled with an electrolyte solution such as KCl to conduct current. Initially, with the microelectrode in the extracellular medium, no potential difference is recorded, simply because the extracellular medium (the extracellular fluid) is isopotential. If, however, the microelectrode penetrates the cell membrane so that the tip of the microelectrode is inside the cell, a sharp deflection is obtained on the recording device (Fig. 8.2B). The potential suddenly shifts from its initial value of 0 mV to a new value of −60 mV. The inside of the cell is negative with respect to the outside. The potential difference that is recorded when a living cell is impaled with a microelectrode is known as the resting potential. The

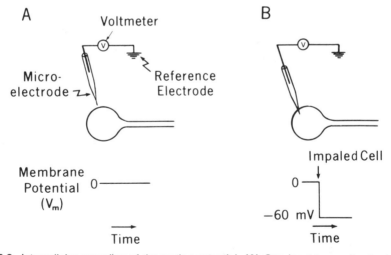

FIG. 8.2. Intracellular recording of the resting potential. **(A)** One input to a voltmeter is connected to a microelectrode, and the second input is connected to a reference electrode in the extracellular medium. No potential difference is recorded when the tip of the microelectrode is outside the cell. **(B)** When the tip of the microelectrode penetrates the cell a resting potential of −60 mV is recorded. (Modified from E. R. Kandel, *The cellular basis of behavior*. San Francisco: Freeman, 1976; Chapters 5 and 6.)

resting potential remains constant for indefinite periods of time as long as the cell is not stimulated or no damage occurs to the cell with impalement. The resting potential varies somewhat from nerve cell to nerve cell (-40 to -90 mV), but a typical value is about -60 mV.

INTRACELLULAR RECORDING OF THE NERVE ACTION POTENTIAL

The techniques for examining resting potentials can be extended to study the action potential. Although nerve action potentials are normally initiated by mechanical, chemical, or photic stimuli to classes of specialized receptors or by a process known as synaptic transmission (see Chapters 12 and 13), it is possible to artificially elicit action potentials in nerve cells and study their underlying ionic mechanisms in considerable detail and in a controlled fashion.

Figure 8.3 shows another idealized nerve cell with its cell body and attached

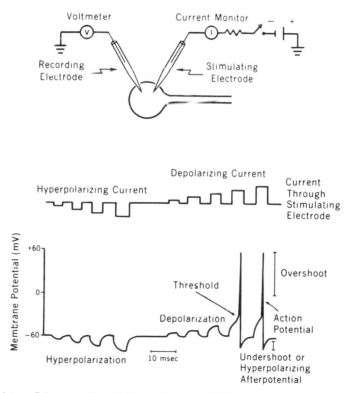

FIG. 8.3. Intracellular recording of the action potential. **Top:** A second intracellular micro-electrode is used to hyperpolarize or depolarize the cell artificially. **Center:** Hyperpolarizing and depolarizing current pulses of increasing amplitudes were passed into the cell. **Bottom:** If the magnitude of the depolarizing current is sufficient to depolarize the membrane potential to threshold, an action potential is initiated. (Modified from E. R. Kandel, *The cellular basis of behavior.* San Francisco: Freeman, 1976; Chapters 5 and 6.)

axon. One microelectrode has penetrated the cell membrane so that the tip of the electrode is inside the cell. This electrode will be used to monitor the potential difference between the outside and inside of the cell. When this electrode penetrates the cell, a resting potential of about -60 mV is recorded. The cell is also impaled with a second microelectrode that will be used to alter the membrane potential artificially. This second electrode, called the stimulating electrode, is connected to a suitable current generator (in the simplest case, this current generator can be considered a battery). Obviously, there are two ways that a battery can be connected to any circuit. The battery can be inserted so that its positive pole is connected to the electrode, or the battery can be inserted so that its negative pole is connected to the electrode. A switch is placed in the circuit so that the battery can be connected and disconnected to the circuit. Assume that a small battery is inserted and its negative pole is connected to the stimulating electrode. With the switch open, a resting potential of -60 mV is recorded. As a result of closing the switch, however, the negative pole of the battery is connected to the stimulating electrode, which tends to artificially make the inside of the cell more negative relative to the external solution. There is a slight downward deflection of the recording trace.

If we repeat this experiment but instead use a slightly larger battery, more current flows into the cell, and a larger increase in the negativity of the cell is recorded. Larger batteries produce even greater increases in the potential. Any time the negativity of the cell interior is increased, the potential change is known as a hyperpolarization. The membrane is more polarized than normal.

This experiment can be repeated in a slightly different way by connecting the positive pole of the battery to the stimulating electrode. Turning on the switch now makes the inside of the cell artificially more positive relative to the external solution. The polarized state of the membrane is decreased. Increasing the size of the battery produces a greater decrease in the negativity of the cell, and over a limited range the resultant potential is a graded function of the size of the stimulus that is used to produce it. Any time the interior of the cell becomes more positive, the potential change is known as a depolarization.

These hyperpolarizations and depolarizations that are artificially produced are known as electrotonic, graded, or passive potentials. Some additional features of electrotonic potentials are discussed later. The point to note here, however, is that, within a limited range of stimulus intensities, hyperpolarizing and depolarizing electrotonic potentials are graded functions of the size of the stimulus used to produce them.

An interesting phenomenon occurs when the magnitude of the battery used to produce the depolarizing potentials is increased further. As the size of the battery and thus the amount of depolarization is increased, a critical level is reached known as the *threshold*, where a new type of potential is produced that is different in its amplitude, duration, and form from the depolarizing pulse used to produce it. This new type of potential change elicited when threshold is reached is known as the action potential. The action potential is elicited in an all-or-nothing fashion. Stimuli below threshold fail to elicit an action potential; stimuli at threshold or above threshold successfully elicit an action potential. Increasing the stimulus intensity beyond threshold produces an action potential that is identical to the action potential produced at the threshold level. In this experiment, the duration of the depolarization is so short that only a single action potential could be initiated. If the duration is longer, multiple action potentials are initiated, and their frequency depends on the stimulus intensity. This is simply a restatement of the all-or-nothing law of action potentials presented earlier. Below threshold, no action potential is elicited; at or above threshold, an all-or-nothing action potential is initiated. Increasing the stimulus intensity still further produces the same amplitude action potential; only the frequency is increased.

Not only are action potentials elicited in an all-or-nothing fashion, but as is described in Chapter 10 they also propagate in an all-or-nothing fashion. If an action potential is initiated in the cell body, it will propagate along the nerve axon and eventually invade the synaptic terminals and initiate a process known as synaptic transmission (see Chapters 12 and 13). Unlike action potentials, electrotonic potentials do not propagate in an all-or-nothing fashion. Electrotonic potentials do spread but only for short distances (see Chapter 10).

There are several interesting features of the action potential. One is that the polarity of the cell completely reverses during the peak of the action potential. Initially, the inside of the cell is -60 mV with respect to the outside, but, during the peak of the action potential, the potential reverses and approaches a value $+55$ mV inside with respect to the outside. The region of the action potential that varies between the 0-mV level and its peak value is known as the *overshoot*. Another interesting characteristic of action potentials is their repolarization phase (the return to the resting level). The action potential does not immediately return to the resting potential of -60 mV; there is a period of time when the cell is actually more negative than the resting level. This phase of the action potential is known as the *undershoot* or the *hyperpolarizing* afterpotential.

As indicated earlier, nerve potentials are the vehicles by which peripheral information is coded and propagated to the central nervous system; motor commands initiated in the central nervous system are propagated to the periphery by nerve action potentials, and the action potential is the first step in the initiation of muscular contraction.

IONIC MECHANISMS OF THE RESTING POTENTIAL

Although the major focus of this chapter and that of Chapter 9 is to explain the ionic mechanisms that underlie the action potential, it is first necessary to review the ionic mechanisms that underlie the resting potential, since the two are intimately related. The basic principles have been introduced in Chapter 7.

Bernstein's Hypothesis (for the Resting Potential)

In 1902, Julius Bernstein proposed the first satisfactory hypothesis for the generation of the resting potential. Bernstein knew that the inside of cells have high K^+

and low Na^+ concentrations, and that the extracellular fluid has low K^+ and high Na^+ concentrations. In addition, there appeared to be large negatively charged molecules, presumably proteins, to which the cell was impermeable. Bernstein also knew (a critical piece of information) that cells were highly permeable to K^+ but not very permeable to other ions. Furthermore, Bernstein knew of the work of the physical chemist, Nernst. Bernstein therefore suggested that one could predict the resting potential simply by applying the Nernst equilibrium equation for potassium:

$$V_m \overset{?}{=} E_K = 60 \log \frac{[K^+]_o}{[K^+]_i} \text{ (mV)} \qquad [8.1]$$

where V_m is the membrane potential and E_K the potassium equilibrium potential (see also Chapter 4).

Although Bernstein's hypothesis was very interesting, it could not be directly tested at the time (hence, the question mark in the equation) because microelectrode recording techniques had not been developed. It was not until the 1930s and 1940s and the advent of microelectrode recording techniques that it became possible to test the hypothesis directly. The testing of Bernstein's hypothesis was done primarily by Hodgkin and Huxley and their colleagues in England. As a result of this work, a general theory was developed for the generation of the resting potential that appears to be applicable to most cells in the body.

Testing Bernstein's Hypothesis

How would one go about testing Bernstein's hypothesis? If the membrane potential (V_m) is equal to the K^+ equilibrium potential (E_K), one should be able to substitute the known outside and inside concentrations of K^+ (Table 8.1) into the Nernst equation and determine the equilibrium potential (E_K), which should equal the measured membrane potential (V_m). Furthermore, because of the logarithmic relationship in the Nernst equation, if the outside K^+ concentration is artificially manipulated by a factor of 10, then the equilibrium potential will change by a factor of 60 mV. If the membrane potential is governed by the K^+ equilibrium potential, then the membrane potential should also change by 60 mV.

TABLE 8.1. *Some common values of ion concentrations*

Ion	Extracellular concentration (mM)	Intracellular concentration (mM)	Nernst potential (mV)
Squid giant axon			
Na^+	440	50	+55
K^+	20	400	−75
Mammalian muscle fiber			
Na^+	145	12	+63
K^+	4	155	−92

Figure 8.4 illustrates one direct experimental test of Bernstein's hypothesis performed by Hodgkin and Horowicz. A cell was impaled with a microelectrode, and the resting potential was measured. The extracellular K^+ concentration was systematically varied, and the change in the resting potential was monitored. When the K^+ concentration was changed by a factor of 10, the resting potential changed by a factor of 60 mV. The straight line on the plot is the relationship predicted by the Nernst equation (note that it is a straight line because the data are plotted on a semilog scale).

The fit is not perfect, however, and the experimental data deviate from the predicted values when the extracellular K^+ concentration is reduced to low levels. If there is a deviation from the Nernst equation, the membrane must be permeable, not only to K^+, but to another ion as well. That other ion appears to be Na^+. As indicated earlier, Na^+ has a high concentration outside the cell and a low concentration inside the cell. If the cell has a slight permeability to Na^+, Na^+ will tend to diffuse into the cell and produce a charge distribution across the membrane so that the inside of the membrane will be positive with respect to the outside. This slight increase in the positivity on the inside surface of the membrane will tend to reduce

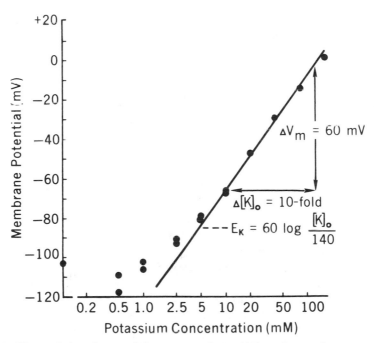

FIG. 8.4. Effects of altered extracellular concentrations of K^+ on the membrane potential: (•) measured membrane potential at each of a variety of different concentrations of K^+; (—) potential predicted by the Nernst equation. The value of 140 in the Nernst equation is the estimated intracellular concentration of K^+ for the cell used in this experiment. (Modified from A. L. Hodgkin and P. Horowicz, *J. Physiol.* 1959;148:127.)

the negative charge distribution produced by the diffusion of K^+ out of the cell. The slight permeability of the membrane to Na^+ will tend, therefore, to make the cell slightly less negative than would be expected were the membrane only permeable to K^+. If a membrane is permeable to more than one cation, the Nernst equation cannot be used to predict the resultant membrane potential. In such a case, however, the Goldman equation can be utilized.

GOLDMAN-HODGKIN-KATZ EQUATION

The Goldman equation is also known as the Goldman-Hodgkin-Katz (GHK) equation because Hodgkin and Katz applied it to biological membranes. As has already been seen in Chapter 7, the GHK equation can be used to determine the potential developed across a membrane permeable to Na^+ and K^+. Thus,

$$V_m = 60 \log \frac{[K^+]_o + \alpha[Na^+]_o}{[K^+]_i + \alpha[Na^+]_i} \text{ (mV)} \qquad [8.2]$$

where V_m is the membrane potential in millivolts and α is equal to the ratio of the Na^+ and K^+ permeabilities (P_{Na}/P_K). This equation looks rather complex at first, but it can be pared down to size by examining two extreme cases. Consider the case when the Na^+ permeability is equal to zero. Then, α is equal to zero, and the GHK equation reduces to the Nernst equation for K^+. If the membrane is highly permeable to Na^+ and has a very low K^+ permeability, α will be a very large number, which causes the Na^+ terms to be very large so that the K^+ terms can be neglected and the GHK equation reduces to the Nernst equation for Na^+. Thus, the GHK equation has two extremes. In one case, when Na^+ permeability is zero, it reduces to the Nernst equation for K^+; in the other case, when Na^+ permeability is very high, it reduces to the Nernst equation for Na^+. The GHK equation allows one to predict membrane potentials between these two extreme levels, and these membrane potentials are determined by the ratio of K^+ and Na^+ permeabilities. If the permeabilities are equal, the membrane potential will be intermediate between the K^+ and the Na^+ equilibrium potentials.

Figure 8.5 illustrates a test of the ability of the GHK equation to fit the same experimental data shown in Fig. 8.4. The straight line is generated by the Nernst equation, whereas the curved trace is generated by the GHK equation. The value of α that gives the best fit is 0.01. Thus, although there is some Na^+ permeability at rest, it is only one hundredth that of the K^+ permeability. To a first approximation, the membrane potential is due to the fact that there is unequal distribution of K^+, and the membrane is selectively permeable to K^+ and to a large extent no other ion. Therefore, the membrane potential can be roughly predicted by the Nernst equilibrium potentials for K^+. However, there is a slight Na^+ permeability that tends to make the inside of the cell more positive than would be predicted, based on the assumption that the cell is only permeable to K^+. The GHK equation can be used to calculate or predict the membrane potential knowing the ratio of Na^+ and K^+

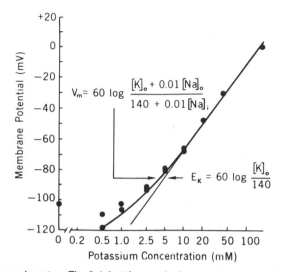

FIG. 8.5. Same experiment as Fig. 8.4, but the graph also contains the prediction of the change in membrane potential obtained with the GHK equation with a value of α equal to 0.01. (Modified from A. L. Hodgkin and P. Horowicz, *J. Physiol.* 1959;148:127.)

permeabilities and the individual extracellular and intracellular concentrations of Na^+ and K^+.

BIBLIOGRAPHY

Aidley DJ. *The physiology of excitable cells*, 2nd ed. Cambridge: Cambridge University Press, 1978; Chapters 1–3.

Hille B. *Ionic channels of excitable membranes*. 2nd Edition, Sunderland, MA: Sinauer Associates, 1992; Chapter 13.

Kandel ER. *The cellular basis of behavior*, San Francisco: Freeman, 1976; Chapters 5 and 6.

Kandel ER, Schwartz JH, Jessell TM. *Principles of neural science*, 3rd ed. New York: Elsevier/North-Holland, 1991; Chapters 2 and 6.

Katz B. *Nerve muscle and synapse*. New York: McGraw-Hill, 1966; Chapters 3 and 4.

Nicholls JG, Martin AR, Wallace BG. *From neuron to brain*, 3rd ed. Sunderland, MA: Sinauer Associates, 1992: Chapters 2 and 3.

Schmidt RF, ed. *Fundamentals of neurophysiology*, 2nd ed. New York: Springer-Verlag, 1978; Chapter 2.

ADDITIONAL READING

Geddes LA. A short history of the electrical stimulation of excitable tissue. *Physiologist* 1984; 27(Suppl):S1–S47.

Hartline HK. Intensity and duration in the excitation of single photoreceptor units. *J Cell Comp Physiol* 1934;5:229.

Hodgkin AL, Horowicz P. The influence of potassium and chloride ions on the membrane potential of single muscle fibres. *J Physiol* 1959;148:127.

Wu CH. Electric fish and the discovery of animal electricity. *Am Sci* 1984;72:598–607.

9

Ionic Mechanisms Underlying the Action Potential

Having reviewed the ionic mechanisms that account for the generation of the resting potential, we can now examine the ionic mechanisms that account for the action potential.

THE SODIUM HYPOTHESIS FOR THE NERVE ACTION POTENTIAL

Is it possible to specify ionic mechanisms that account for the action potential just as it was possible to do so for the resting potential? It is interesting to note that Julius Bernstein in 1902, when proposing his theory for the resting potential, also proposed a theory for the nerve action potential. Bernstein proposed that during a nerve action potential, the membrane suddenly became permeable to all ions. Bernstein predicted, based on this theory, that the membrane potential would shift from its resting level to a new value of about 0 mV. (Can you explain why?) We have already learned, however, that the potential changes during the action potential do not range from a value of -60 to 0 mV, but actually go well beyond 0 mV and approach a value of $+55$ mV. So while Bernstein's hypothesis for the resting potential was nearly correct, his hypothesis for the action potential clearly missed the mark.

At the same time that Bernstein proposed theories for resting potentials and action potentials, Overton, another physiologist, made some interesting observations about the critical role of Na^+. Overton observed that Na^+ in the extracellular medium was absolutely essential for cellular excitability. In general, in the absence of extracellular Na^+, nerve axons cannot propagate information, and *skeletal* muscle cells are unable to contract.

Overton, like Bernstein, could not test his hypothesis experimentally because microelectrodes were not available. Just as Hodgkin and his colleagues critically tested Bernstein's hypothesis for the resting potential, they also examined and extended Overton's observations. One of the earlier experiments performed by Hodgkin and Katz is illustrated in Fig. 9.1.

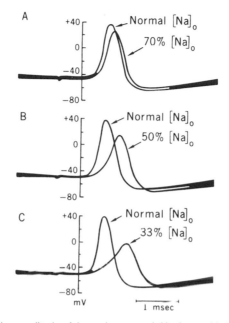

FIG. 9.1. Changes in the amplitude of the action potential in the squid giant axon as a function of the extracellular concentration of Na^+ reduced to 70% **(A)**, 50% **(B)**, and 33% **(C)** of its normal value. (Modified from A. L. Hodgkin and B. Katz, *J Physiol* 1949;108.37.)

Hodgkin and Katz repeatedly initiated action potentials in the squid giant axon while they artificially altered the extracellular Na^+ concentration. When the extracellular Na^+ concentration was reduced to 70% of its normal value (Fig. 9.1A), there was a slight reduction in the amplitude of the action potential. Reducing the Na^+ concentration to 50% and 33% of its normal value produced further reductions in the amplitude of the action potential. These experiments, therefore, directly confirmed Overton's initial observations that Na^+ is essential for the initiation of action potentials (exceptions to this are action potentials in cardiac and smooth muscle cells).

Hodgkin and his colleagues took these observations one step further. They suggested that during an action potential, the membrane behaved as though it was becoming selectively permeable to Na^+. In a sense, the membrane was "switching" from its state of being highly permeable to K^+ at rest to being highly permeable to Na^+ at the peak of the action potential. If a membrane is highly permeable to Na^+ at the peak of the action potential (for the sake of simplicity we assume that the membrane is solely permeable to Na^+ and no other ions), what potential difference would one predict across the cell membrane? If the membrane is only permeable to Na^+, the membrane potential should equal the Na^+ equilibrium potential (E_{Na}); and

$$V_m \overset{?}{=} E_{Na} = 60 \log \frac{[Na^+]_o}{[Na^+]_i} \ (mV) \tag{9.1}$$

Indeed, when the known values of extracellular and intracellular Na^+ concentrations for the squid giant axon are substituted, a value of $+55$ mV is calculated. This is approximately the peak amplitude of the action potential. Is this simply a coincidence? It is possible that the membrane is permeable to other ions as well. Perhaps the action potential is due to an increase in Ca^{2+} permeability; Ca^{2+} is in high concentration outside and low concentration inside the cell, so part of the action potential might be due to a selective increase in Ca^{2+} permeability. How can this issue be resolved? If the peak amplitude of the action potential is determined by E_{Na}, one would expect that as the extracellular levels of Na^+ are altered, the peak amplitude of the action potential would change according to the Nernst equation. Furthermore, because of the logarithmic relationship in the Nernst equation, if the extracellular Na^+ concentration is changed by a factor of 10, the Na^+ equilibrium potential and the peak amplitude of the action potential should change by a factor of 60 mV.

Figure 9.2 illustrates a test of this hypothesis. The peak amplitude of the action potential, shown on the vertical axis, is measured as a function of the extracellular Na^+ concentration. The dots on the graph represent the peak amplitude of the action potential recorded at various extracellular concentrations of Na^+. The straight line

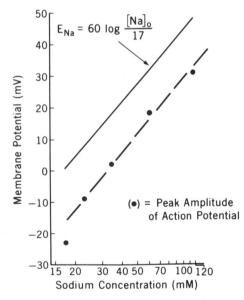

FIG. 9.2. Plot of the peak amplitude of the action potential vs. the extracellular concentration of Na^+. The *solid line* is a prediction of the Nernst equation with an estimated intracellular Na^+ concentration for this cell of 17 mM. (Modified from W. L. Nastuk and A. L. Hodgkin *J Cell Comp Physiol* 1950;35:39.)

is the relationship that describes the Na^+ equilibrium potential as a function of extracellular Na^+.

Although there are some deviations between the predicted Na^+ equilibrium potential and the peak amplitude of the action potential (the action potential never quite reaches the value of the Na^+ equilibrium potential), the critical observation is that the slopes of these two lines are nearly identical. For a tenfold change in the extracellular Na^+ concentration, there is approximately a 60-mV change in the peak amplitude of the action potential. These experiments, therefore, provide strong experimental support for the hypothesis that during the peak of the action potential, the membrane suddenly switches from a high permeability to K^+ to a high permeability to Na^+.

APPLYING THE GHK EQUATION TO THE ACTION POTENTIAL

There are two important positively charged ions (K^+ and Na^+), and the membrane potential appears to be governed by the relative permeabilities of these two ions. As a result, the GHK equation can be utilized:

$$V_m = 60 \log \frac{[K^+]_o + \alpha[Na^+]_o}{[K^+]_i + \alpha[Na^+]_i} \ (mV)$$ [9.2]

where $\alpha = P_{Na}/P_K$.

Figure 9.3 is a sketch of an action potential. One important observation is that the action potential traverses a region that is bounded by E_{Na} on one extreme and E_K on the other. Because the action potential traverses this bounded region, it is possible to utilize the GHK equation to predict any value of the action potential simply by adjusting the ratio of the Na^+ and K^+ permeabilities. For the resting level, we have already seen that the ratio of Na^+ and K^+ permeabilities is 0.01. Thus, we can

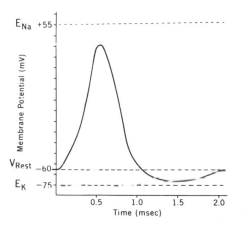

FIG. 9.3. Sketch of a nerve action potential.

substitute these values into the GHK equation and calculate a value of approximately -60 mV.

Assume that the Na^+ permeability is very high. Then α is a very large number, and the Na^+ terms dominate the GHK equation. In the limit, the GHK equation reduces to the Nernst equation for Na^+. So, when there is a high Na^+ permeability and a low K^+ permeability, we can calculate a potential that approximates the peak amplitude of the action potential. During the repolarization phase of the action potential, we can simply assume that the ratio of Na^+ and K^+ permeabilities returns back to normal, substitute this value into the GHK equation and calculate a membrane potential of -60 mV. The hyperpolarizing afterpotential could be accounted for by a slight decrease in Na^+ permeability to less than its resting level or by a K^+ permeability greater than its resting level.

The important point is that by adjusting the ratio of Na^+ and K^+ permeabilities, it is possible to predict the entire trajectory of the action potential.

THE CONCEPT OF VOLTAGE-DEPENDENT Na^+ PERMEABILITY

Despite the fact that the GHK equation gives a good qualitative fit to the trajectory of the action potential, it fails to provide any insight into the fundamental question of how the presumed switch in permeability takes place. How can a membrane at one instant in time be highly permeable to K^+ and a short time thereafter be highly permeable to Na^+? Hodgkin and Huxley proposed that there is a voltage-dependent change in Na^+ permeability; Na^+ permeability is low at rest, but as the cell is depolarized, Na^+ permeability increases (Fig. 9.4A).

Assume that the cell is depolarized by some stimulus. As a result of the depolarization, there will be an increase in Na^+ permeability. If Na^+ permeability is in-

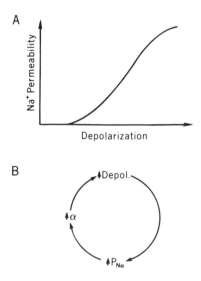

FIG. 9.4. Relationships between depolarization and Na^+ permeability critical for the initiation of an action potential (see text for explanation).

creased (assume for the moment that K^+ permeability remains unchanged), α in the GHK equation will be increased. If α is increased, the Na^+ terms are multiplied by a larger value, and they will tend to dominate the GHK equation. The membrane will become less negative (more depolarized); but as the cell depolarizes, Na^+ permeability increases further and α increases further. A positive feedback cycle is entered (Fig. 9.4B) such that once the cell is depolarized to a critical level, the cell will rapidly depolarize further in a regenerative fashion. Eventually, the membrane potential will approach the Na^+ equilibrium potential. Thus, a voltage-dependent relationship between membrane potential and Na^+ permeability can in principle completely account for the initiation of the action potential.

TESTING THE CONCEPT OF VOLTAGE-DEPENDENT Na^+ PERMEABILITY: THE VOLTAGE CLAMP TECHNIQUE

So far we have just a theory. The critical hypothesis is that Na^+ permeability is regulated by the membrane potential. The simple way of testing this hypothesis is to depolarize the cell to various levels and measure the corresponding Na^+ permeability. The problem, however, is that as soon as the cell is depolarized, Na^+ permeability changes, an action potential is initiated, and due to practical reasons, there is insufficient time to measure the permeability change. This was a major obstacle in the further analysis of the ionic mechanisms that govern the action potential.

Hodgkin and Huxley and their colleagues devised a scheme that allowed them to stabilize the membrane potential at various levels for indefinite periods of time. They used an electronic feedback device known as a voltage clamp amplifier to hold the membrane potential at various levels (Fig. 9.5). The voltage clamp amplifier takes the difference between the actual recorded membrane potential and the desired

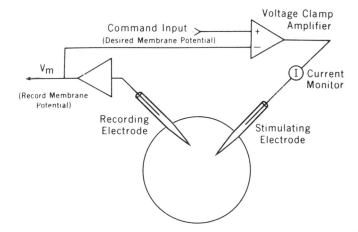

FIG. 9.5. Schematic diagram of the voltage clamp apparatus (see text for details).

level and generates sufficient hyperpolarizing or depolarizing current to minimize the difference. The amount of current necessary to hold the membrane potential fixed at the desired level provides an index of the membrane permeability, or conductance, at that particular voltage clamp level. For example, by measuring the ionic current as a function of time, $I(t)$, and knowing the potential difference (which is constant), the conductance as a function of time, $G(t)$, can be determined simply by using Ohm's law (conductance for our purpose can simply be considered an electrical measurement of permeability, so we will use permeability and conductance interchangeably):

$$G(t) = I(t)/\Delta V \qquad\qquad [9.3]$$

By changing the potential difference with the voltage clamp amplifier, the corresponding conductances at a variety of different potentials can be determined.

Figure 9.6 illustrates some typical results. The procedure is as follows. Initially, the membrane potential is at its resting level of -60 mV. It is then artificially changed from the resting level to a new depolarized level (e.g., -35 mV) and held there for 5 msec or longer. The membrane potential is then returned back to its resting level, and the membrane is stepped or clamped to a new depolarized level of

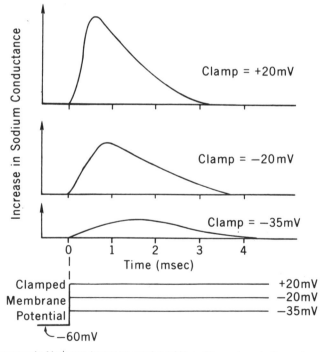

FIG. 9.6. Changes in Na^+ conductance produced by voltage steps to three depolarized levels. The greater the depolarization, the greater the amplitude of Na^+ conductance. (Modified from A. L. Hodgkin and A. F. Huxley, *J Physiol* 1952;117:500.)

−20 mV. By performing a sequence of these voltage clamp measurements, changes in Na^+ permeability as a function of both voltage and time can be determined.

In the upper part of Fig. 9.6, the horizontal axis shows the time and the vertical axis shows the measured Na^+ conductance. As the membrane potential is forced to various depolarized levels from the resting level, there is a graded increase in Na^+ permeability. The greater the level of depolarization, the greater the Na^+ permeability. This experiment therefore provides strong experimental support for the proposal that Na^+ permeability is voltage-dependent and demonstrates the existence of a mechanism that could explain the rising phase (initiation) of the action potential.

MOLECULAR BASIS FOR THE REGULATION OF Na^+ PERMEABILITY

At the molecular level, the relationship between the membrane potential and Na^+ permeability (Fig. 9.4A) is due to the existence of membrane channels that are selectively permeable to Na^+ and that are opened or gated by the membrane potential. This discovery was made possible by the patch clamp technique that allows the conductance of individual channels to be measured. With the patch clamp technique, a micropipette with a tip several microns in diameter is positioned so that the tip just touches the outer surface of the membrane (see Chapter 5). A high-resistance seal develops that allows the electrode and associated electronic circuitry to measure the current and thus the conductance of a small number of Na^+ channels or, indeed, a single Na^+ channel (Fig. 9.7). One of the major conclusions that has been derived from these studies is that, in response to membrane depolarization, single Na^+ channels open in all-or-nothing fashion. A single channel has at least two states—open and closed—and once opened it cannot open further in response to depolarization. The gating process is subserved by a membrane-bound protein that is charged, such that when the membrane is depolarized, a conformational change in the protein takes place that results in the channel becoming more permeable to Na^+ (see also below).

Individual channels open briefly and then close (Fig. 9.8). The opening of single channels is a probabilistic function of time and voltage; however, when the opening of many channels is averaged, the averaged conductance predicts the conductance change of the entire population of channels (Fig. 9.9). Thus, the time course of the changes of Na^+ permeability (Fig. 9.6) is a reflection of the average opening and closing times of many individual Na^+ channels. At the molecular level, the voltage dependence of the total membrane Na^+ permeability (Fig. 9.4A) can be viewed as the probability that a depolarization will open single Na^+ channels; the more the cell is depolarized, the greater the number of individual Na^+ channels that will be opened each in its characteristic all-or-nothing fashion.

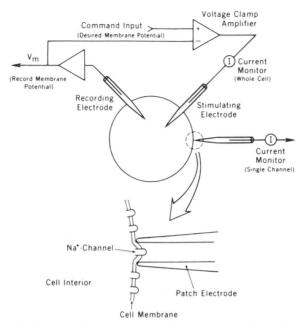

FIG. 9.7. Schematic diagram of the apparatus used for single-channel recording. A micropipette is positioned to touch the surface of the membrane. A tight seal develops, and the current flowing through individual channels can be monitored. Membrane potential can be altered by the conventional two-microelectrode technique.

STRUCTURE OF THE VOLTAGE-GATED Na$^+$ CHANNEL

The principal structural and functional unit of the voltage-gated Na$^+$ channel consists of a single polypeptide chain exhibiting four homologous domains (I to IV, Fig. 9.9), with each domain having six hydrophobic membrane-spanning regions designated S1 to S6. The functional significance of specific regions is now being elucidated. For example, region S4 contains a high density of positively charged residues and is believed to represent the channel's voltage sensor. The channel pore is believed to be formed by the four homologous regions between S5 and S6 (also designated the SS1 and SS2 regions, the H5 region or the "P" region). Channel inactivation (see below) seems to be associated with the region that links domains III and IV. In particular, a hinged-lid structure formed by the amino terminus of domain IV has properties consistent with an ability to move into the channel pore and cause its blockade. Finally, the Na$^+$ channel can be regulated by protein phosphorylation. Specifically, the region that links domains I and II contains phosphorylation sites for cAMP-dependent protein kinase, whereas the region that links domains III and IV contains a phosphorylation site for protein kinase C.

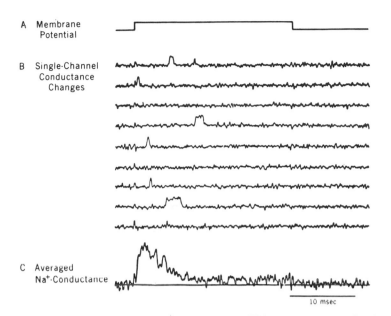

FIG. 9.8. Single-channel changes in Na$^+$ conductance. **(A)** In response to a pulse depolarization, the probability of single Na$^+$ channels opening is increased. **(B)** Successive traces obtained in response to multiple presentations of the voltage step in **(A)**. With any single trace the relationship is not particularly clear. **(C)** When the traces are averaged, a conductance change resembling that of the macroscopic conductance change is observed. (Modified from F. J. Sigworth and E. Neher, *Nature* 1980;287:447–449.)

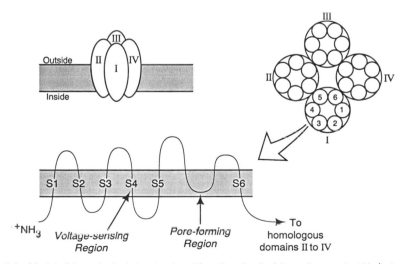

FIG. 9.9. Model of the principal structural and functional unit of the voltage-gated Na$^+$ channel.

Na$^+$ INACTIVATION

There is one other important aspect of the data illustrated in Fig. 9.6. Despite the fact that the membrane potential is depolarized throughout the duration of each trace, Na$^+$ permeability spontaneously falls back to its resting level. Thus, although Na$^+$ permeability is dependent on the level of depolarization, it does not remain elevated and is only transient. Once it reaches its maximum value, it spontaneously decays back to its resting level. The process by which Na$^+$ permeability spontaneously decays back to its resting level (despite the fact that the membrane is depolarized) is known as *inactivation*. At the molecular level, the process of inactivation can be considered to be a separate voltage- and time-dependent process regulating the Na$^+$ channel. In Fig. 9.10, the Na$^+$ channel is represented as having two regulatory components: an activation gate and an inactivation gate. For the channel to be open, both the activation and inactivation gates must be open. At the resting potential, the activation gate is closed, and despite the fact that the inactivation gate is open, channel permeability is zero (Fig. 9.10A). With depolarization, the activation gate opens rapidly (Fig. 9.10B), and the channel becomes permeable to Na$^+$. Depolarization also tends to close the inactivation gate, but the inactivation process is slower. With depolarization occurring over a longer time, the inactivation gate closes, and even though the activation gate is still open, channel permeability is zero (Fig. 9.10C).

What is the physiological significance of Na$^+$ inactivation? Let us return to the positive feedback cycle once again. Depolarization increases Na$^+$ permeability, and the increase in Na$^+$ permeability depolarizes the cell. Eventually, as a result of this regenerative cycle, the cell is rapidly depolarized up to a value near E_{Na}. The problem with this mechanism is how to account for the repolarization phase of the action potential. Based on the relationship between Na$^+$ permeability and membrane potential, one would predict that once the membrane potential moves to E_{Na} it would stay there for an indefinite period of time. The steep relationship between voltage and Na$^+$ permeability is only transient, however. After approximately 1 msec, Na$^+$ permeability spontaneously decays. If Na$^+$ permeability decays, due to inactivation, the potential would move closer to E_K; or, stated in a slightly different way, it will become less depolarized. Depolarization would be reduced and the reduction in depolarization would produce a further reduction in Na$^+$ permeability because of the basic relationship between voltage and Na$^+$ permeability (Fig. 9.4A). As a result, a new feedback cycle is initiated that would tend to repolarize the cell.

It is therefore intriguing to think that simply by accounting for (a) the voltage-dependent increase in Na$^+$ permeability and (b) the process of Na$^+$ inactivation, both the initiation and the repolarization phases of the action potential could be explained fully. There are at least two problems with this hypothesis. First, the duration of the action potential is only about 1 msec, yet the Na$^+$ permeability change takes 4 msec or so to return to its resting level (fully inactivate). So by extrapolating these voltage clamp measurements, one might expect that the action potential would be somewhat longer in duration than 1 msec. Second, it is difficult

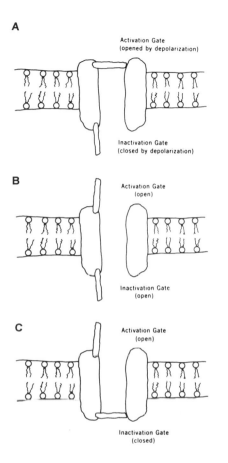

FIG. 9.10. Schematic diagram of three states of the Na$^+$ channel (see text for details). **(A)** Rest. **(B)** Peak g_{Na}. **(C)** Inactivated.

to explain the hyperpolarizing afterpotential. From the voltage clamp data (Fig. 9.6) it is clear that Na$^+$ permeability increases dramatically and then inactivates. To explain the hyperpolarizing afterpotential, Na$^+$ permeability would have to be less than its initial value (α would have to be less than 0.01). So, based on the observed changes in Na$^+$ permeability, it would be impossible to account for the hyperpolarizing afterpotential.

ROLE OF VOLTAGE-DEPENDENT K$^+$ CONDUCTANCE IN THE REPOLARIZATION OF THE ACTION POTENTIAL

Not only does Na$^+$ permeability change, but there is also a change in K$^+$ permeability during the course of an action potential. Figure 9.11 illustrates these results. The upper portion of the illustration simply reviews the experimental data of Fig. 9.6. When the membrane is depolarized and held fixed at various levels, there is an increase in Na$^+$ permeability that is proportional to the depolarization. Keep in mind that during this entire sequence of events, the membrane potential is held depolarized by the voltage clamp at the levels indicated. The traces below (Fig. 9.11B) illustrate that in addition to changes of Na^+ permeability, there are also voltage-dependent changes in K^+ permeability. The greater the level of depolarization, the greater the increase in K^+ permeability. There are two important differences between these two permeability systems. First, the changes in K^+ permeability are rather slow. It takes some time for K^+ permeability to begin to increase, whereas the changes in Na^+ permeability begin to occur immediately after the depolarization is delivered. Second, whereas Na^+ permeability exhibits inactivation, K^+ permeability remains elevated as long as the membrane potential is held depolarized.

Now that it is clear that there are changes in both Na^+ and K^+ permeabilities, how can this information be utilized to better account for the entire sequence of events that underlies the action potential? Because of the slowness of K^+ permeability changes, the initial explanation for the rising phase of the action potential

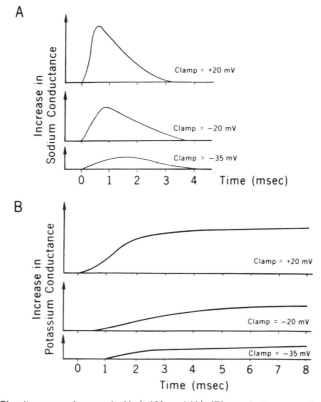

FIG. 9.11. Simultaneous changes in Na^+ **(A)** and K^+ **(B)** conductance produced by voltage steps to three depolarized levels. Note the marked differences between the changes in Na^+ and K^+ conductance. (Modified from A. L. Hodgkin and A. F. Huxley, *J Physiol* 1952;117:500.)

is unaltered, simply because for a period of time less than about 0.5 msec, there is no major change in K^+ permeability. In later phases of the action potential (at times greater than roughly 0.5–1 msec), we not only have to consider Na^+ permeability changes but also changes in K^+ permeability. What would be the consequences of not only having a fall in Na^+ permeability (due to inactivation) but also a simultaneous increase in K^+ permeability? Let us return to the GHK equation. At the peak of the action potential (about 0.5 to 1 msec from its initiation), there is very high Na^+ permeability. At this time, K^+ permeability begins to increase significantly. Thus, at any time after about 0.5 to 1 msec, not only will there be a certain increase in Na^+ permeability, but there will also be a K^+ permeability that is greater than its resting level. As a result, the value of α will be smaller than if only changes in Na^+ permeability were occurring. If α is smaller, then the Na^+ terms make less of a contribution to the GHK equation. Stated in a slightly different way, the K^+ terms make more of a contribution, and the membrane potential will be more negative. Thus, by incorporating the finding that there is a delayed increase in K^+ permeability, the membrane potential will be more negative for any given time (greater than about 0.5–1 msec) than it would have been without the changes in K^+ permeability. The delayed changes in K^+ permeability will tend to make the membrane potential repolarize faster because now there are two driving forces for repolarization. The first is Na^+ inactivation, and the second is the delayed increase in K^+ permeability. By incorporating the simultaneous changes in K^+ permeability, we can in principle account for a shorter duration action potential.

Can the changes in K^+ permeability help explain the hyperpolarizing afterpotential? The key is understanding the time course of the changes in K^+ permeability. You will note that the changes in K^+ permeability are very slow in turning on. They are also slow in turning off. As the action potential repolarizes to the resting level, Na^+ permeability returns back to its resting level. Because the K^+ permeability system is slow, however, K^+ permeability is still elevated. Therefore, α in the GHK equation will actually be less than its initial level of 0.01. If α is less than 0.01, the contributions of the Na^+ terms become even more negligible than at rest, and the membrane potential approaches E_K. Thus, because Na^+ permeability decays rapidly and K^+ permeability decays slowly, during the later phases of the action potential K^+ permeability is elevated, and the hyperpolarizing afterpotential is produced.

Up to this point, the arguments concerning the sequence of events underlying the action potential have been somewhat qualitative and are based on extrapolation of the voltage clamp data. This approach is somewhat unsatisfactory, and it was also unsatisfactory to Hodgkin and Huxley. Hodgkin and Huxley sought to test more rigorously this hypothesis, and they developed a quantitative mathematical model of the action potential based on the experimentally measured changes in Na^+ and K^+ permeabilities (for details, see the Appendix). They sought to determine whether it was possible to reconstruct an action potential from data based entirely on the changes in Na^+ and K^+ permeabilities that were measured with the voltage clamp technique. Their results are illustrated in Fig. 9.12. The solid line illustrates the

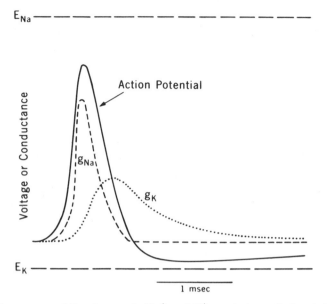

FIG. 9.12. Time course of the changes in Na$^+$ and K$^+$ conductance that underlie the nerve action potential.

computed action potential. This simulated action potential is identical to the experimentally recorded action potential. Since the simulated and experimental action potentials are identical, one can examine in detail the sequence of permeability changes that are necessary to generate the simulated action potential. The dashed line shows the underlying changes in Na$^+$ permeability, and the dotted line shows the underlying changes in K$^+$ permeability.

Assume that by some mechanism the cell is depolarized to threshold. The depolarization initiates the voltage-dependent increase in Na$^+$ permeability. That voltage-dependent increase in Na$^+$ permeability produces a further depolarization that produces further increases in Na$^+$ permeability. The positive-feedback cycle is entered, which leads to a rapid depolarization of the cell toward E_{Na}. At the peak of the spike, which occurs about $\frac{3}{4}$ msec from the initiation of the action potential, two important processes contribute to the repolarization. First, there is the process of Na$^+$ inactivation. As a result of the decay of Na$^+$ permeability, the membrane potential begins to return to the resting level. As the membrane potential returns to the resting level, the Na$^+$ permeability decreases further, which further speeds the repolarization process. A new feedback cycle is entered that moves the membrane potential in the reverse direction. Second, there is the delayed increase in K$^+$ permeability. As the action potential reaches its peak value, there is a rather dramatic change in K$^+$ permeability. This change in K$^+$ permeability tends to move the membrane potential toward E_K. Therefore, there are two independent processes that contribute to repolarization of the action potential. One is Na$^+$ inactivation, and the other is the delayed increase in K^+ permeability. Note that when the action potential returns to its resting level of about -60 mV or so, the Na^+ permeability has reached its resting level; while Na^+ permeability has returned to its resting level, K^+ permeability remains elevated for a period of time. Thus, the ratio of the two permeabilities will be less than it was initially, and the membrane potential will move closer to E_K. Over a period of time, K^+ permeability gradually decays back to its resting level, and the action potential terminates.

In summary, the initiation of the action potential can be explained by the voltage-dependent increase in Na^+ permeability and the repolarization phase of the action potential by (a) the process of Na^+ inactivation and (b) by the delayed increase in K^+ permeability. Finally, the hyperpolarizing afterpotential can be explained by the fact that K^+ permeability remains elevated for a period of time after the Na^+ permeability has returned to its resting level.

Students frequently question the necessity for such an elaborate series of steps to generate short-duration action potentials. This question brings us back to a point raised at the beginning of Chapter 8. Recall that the nervous system codes information in terms of the number of action potentials elicited; the greater the stimulus intensity, the greater the frequency of action potentials. To encode and transmit more information per unit time, it is desirable to generate action potentials at a high frequency. With short-duration action potentials, a new action potential can be initiated soon after the first, and this requirement can be met.

The analysis of Hodgkin and Huxley, originally performed on the squid giant axon, has proved generally applicable to action potentials that are initiated in nerve axons and in skeletal muscle cells. The concept of voltage-dependent ion channels is now universal. What varies from cell to cell is the particular ion to which the channel is permeable. For example, a significant component of action potentials in smooth muscle and cardiac muscle cells is due to voltage-dependent Ca^{2+} channels. A number of different types of voltage-dependent K^+ channels have also been described.

The structure of voltage-gated Ca^{2+} channels is very similar to that of the Na^+ channel (see Fig. 9.9). Voltage-activated K^+ channels have similar six membrane spanning regions, but they differ in that the polypeptide chain does not contain multiply repeated domains. Rather, K^+ channels are formed by the functional association of four separate subunits to form an ionophore. A specific region of the N-terminal domain of the channel peptide appears to be essential for the proper aggregation of the subunits to form the tetrameric structures of the functional channel. Variations in the structure of individual subunits as well as different combinations of the subunits contribute to the great diversity of K^+ channel properties that has been observed in excitable membranes. For example, some K^+ channels exhibit inactivation like Na^+ channels. For these inactivating K^+ channels, the amino terminal sequence of the polypeptide appears to act as a plug to close the channel. Another important class of K^+ channels is activated by intracellular levels of Ca^{2+}. A potential Ca^{2+}-binding domain is found on the carboxy terminal end of the channel polypeptide.

SPECIFICITY OF ION CHANNELS UNDERLYING
THE ACTION POTENTIAL

The voltage clamp analysis and mathematical reconstruction of the action potential leaves many students somewhat unsatisfied because of the seemingly obscure nature of the analysis. Fortunately, there are other types of experiments that confirm further the Hodgkin-Huxley hypothesis for the initiation and repolarization phases of the action potential. Compounds have been discovered that can be used selectively to block or inhibit these voltage-dependent permeability changes. One of these substances is known as tetrodotoxin (TTX), and the other is known as tetraethylammonium (TEA). TTX is a toxin that is isolated from the ovaries of the Japanese puffer fish. Every year in Japan this substance accounts for several hundred deaths due to improper food handling.

The reason why TTX has such devastating effects is illustrated in Fig. 9.13. Shown are the results of a voltage clamp experiment where Na^+ and K^+ per-

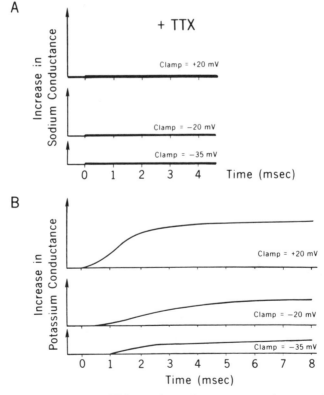

FIG. 9.13. Effects of tetrodotoxin (TTX) on voltage clamp responses from a squid giant axon. **(A)** Na^+ permeability. **(B)** K^+ permeability.

meabilities are measured in the presence of TTX. As a result of perfusing the extra-cellular medium with TTX, the voltage-dependent changes in Na^+ permeability are completely abolished. In contrast, the voltage-dependent changes in K^+ per-meability are unaffected. Figure 9.14 illustrates the results of perfusing a squid giant axon preparation with TEA. TEA has absolutely no effect on the voltage-dependent changes in Na^+ permeability but completely abolishes the voltage-de-pendent changes in K^+ permeability. Thus, one substance (TTX) is capable of blocking the voltage-dependent Na^+ permeability, and another (TEA) is capable of blocking the voltage-dependent K^+ permeability.

Given that the effects of TEA and TTX on permeabilities measured with the voltage clamp are known, how would one expect these substances to affect the action potential? If the voltage-dependent change in Na^+ permeability is blocked, one would expect that no action potential could be initiated or propagated. If the voltage-dependent change in K^+ permeability is blocked, one would expect the action potential to be somewhat longer in duration, and, in addition, it should not

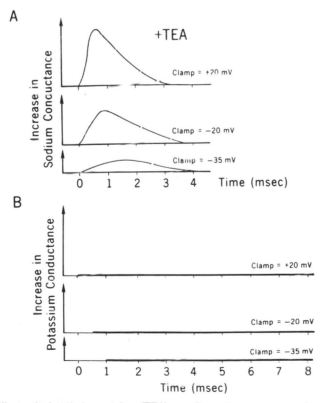

FIG. 9.14. Effects of tetraethylammonium (TEA) on voltage clamp responses from a squid giant axon. **(A)** Na^+ permeability. **(B)** K^+ permeability.

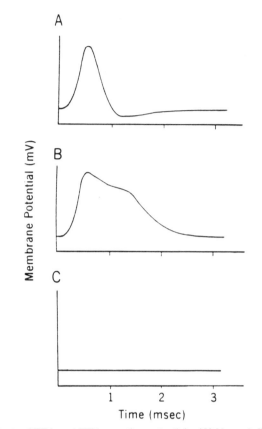

FIG. 9.15. Effects of TEA and TTX on action potentials. **(A)** Normal. **(B)** TEA. **(C)** TTX.

have a hyperpolarizing afterpotential. Figure 9.15 illustrates these results. Figure 9.15A is a normal action potential, and Fig. 9.15B is an action potential recorded in TEA. In TEA, the initiation and the rising phase of the action potential as well as its peak value are unaffected, but there is a dramatic increase in the spike duration and an absence of the hyperpolarizing afterpotential. Thus, the use of TEA confirms the Hodgkin-Huxley theory that the delayed increase in K^+ permeability contributes to the repolarization phase of the action potential and to the undershoot. In the presence of TEA, the process of Na^+ inactivation accounts entirely for the repolarization. Figure 9.15C illustrates the effects of perfusing the preparation with TTX. First, it should be mentioned that in this particular experiment an action potential is elicited in one portion of the axon and is allowed to propagate along the axon to a more distant point where the action potential is recorded. When one perfuses the axon with TTX, one finds that no action potential can be elicited and no action potential can be propagated and monitored at the recording site (Fig. 9.15C). Thus, the use of TTX confirms the Hodgkin-Huxley theory for the critical role that the increase in Na^+ permeability plays in initiating the action potential in nerve axons.

These experiments with TTX and TEA are also interesting in another respect because they demonstrate that the voltage-dependent changes in Na^+ permeability are mediated by completely different membrane channels from the voltage-dependent change in K^+ permeability simply because it is possible to selectively block one but not the other.

DO CHANGES IN Na^+ AND K^+ CONCENTRATIONS OCCUR DURING ACTION POTENTIALS?

It is clear that, as a result of the voltage-dependent increase in Na^+ permeability, some Na^+ ions will flow from the outside of the cell to the inside of the cell. A frequently asked question is whether the flux of Na^+ that occurs during the action potential produces a concentration change on the inside of the cell. Alternatively, since there is an increase in K^+ permeability during an action potential, there is a tendency for some K^+ to flow out of the cell. Does that flux of K^+ cause a change in the intracellular K^+ concentration? Although some Na^+ does indeed enter the cell with each action potential and some K^+ leaves the cell with each action potential, for cells with small surface area-to-volume ratios, these fluxes are generally minute compared to the normal intracellular concentrations. For example, as a result of an action potential, the change in Na^+ concentration for a 1-cm^2 surface area of the membrane is only equal to approximately 1 pM (1×10^{-12} M) and that concentration change is restricted to the inner surface of the membrane. Therefore, despite the fact that some Na^+ does enter the cell with each action potential, the concentration change is minute compared to the normal millimolar concentration of Na^+ within the cell. There is also some K^+ that leaves the cell during an action potential, but the concentration change again is minute compared to the normal K^+ concentration. Indeed, if the (Na^+–K^+) exchange pump is blocked in the squid giant axon, it is possible to initiate more than 500,000 action potentials without any noticeable change in either the resting potential or the amplitude of the action potential.

The role of the membrane (Na^+–K^+) pump is to provide long-term maintenance of the Na^+ and K^+ concentration differences. Eventually, if one were to generate more than 500,000 action potentials, there would be a change in ionic distribution, but this is a long-term phenomenon, and in the short range the (Na^+–K^+) pump is not essential. Even with the (Na^+–K^+) exchange pump blocked, a cell is capable of initiating a large number of action potentials without any major change in either the resting potential or the peak amplitude of the action potential. If the resting potential is unchanged, it can be inferred that the K^+ equilibrium potential is also unchanged. Similarly, if the peak amplitude of the action potential is unchanged, the Na^+ equilibrium potential is also unchanged.

THRESHOLD, ACCOMMODATION, AND ABSOLUTE
AND RELATIVE REFRACTORY PERIODS

The Hodgkin-Huxley analysis not only described quantitatively the mechanisms that account for the initiation and the repolarization of the action potential but also provided the explanation for some phenomena that had been known for some time but were poorly understood. Four of these phenomena are *threshold, accommodation*, and the *absolute* and *relative refractory periods*.

Threshold

The voltage dependence of Na^+ permeability explains the initiation of the action potential but by itself does not explain completely the threshold phenomenon, since the relationship between depolarization and Na^+ permeability, although steep, is a continuous function of membrane depolarization (e.g., Fig. 9.4A). Threshold can be explained by taking into account the fact that K^+ permeability (both resting and voltage-dependent) tends to oppose the effects of increasing Na^+ conductance in depolarizing the cell and initiating an action potential. Threshold is the point where the depolarizing effects of the increased Na^+ permeability just exceed the counter (hyperpolarizing) effects of K^+ permeability. Once the inward flow of Na^+ exceeds the outward flow of K^+, threshold is reached, and an action potential occurs through the positive feedback cycle.

Absolute and Relative Refractory Periods

The absolute refractory period refers to that period of time after the initiation of one action potential when it is impossible to initiate another action potential despite the stimulus intensity utilized. The relative refractory period refers to that period of time after the initiation of one action potential when it is possible to initiate another action potential but only with a stimulus intensity greater than that utilized to produce the first action potential.

At least part of the relative refractory period can be explained by the hyperpolarizing afterpotential. Assume that a cell has a resting potential of -60 mV and a threshold of -45 mV. If the cell is depolarized by 15 mV to reach threshold, an all-or-nothing action potential will be initiated, followed by the associated repolarization phase and the hyperpolarizing afterpotential. What happens if one attempts to initiate a second action potential during the undershoot? Initially the cell was depolarized by 15 mV (from -60 to -45 mV). If the same depolarization (15 mV) is delivered during some phase of the hyperpolarizing afterpotential, the 15 mV depolarization would fail to reach threshold (-45 mV) and initiate an action potential. If, however, the cell is depolarized by more than 15 mV, threshold can again be reached, and another action potential initiated. Eventually, the hyperpolarizing afterpotential would terminate, and the original 15-mV stimulus would again be sufficient to reach threshold. The process of Na$^+$ inactivation also contributes to the relative refractory period (see below).

The absolute refractory period refers to that period of time after an action potential when it is impossible to initiate a new action potential no matter how large the stimulus. This is a relatively short period of time that varies from cell to cell but roughly occurs approximately $\frac{1}{2}$ to 1 msec after the peak of the action potential. To understand the absolute refractory period, it is necessary to understand Na$^+$ inactivation in greater detail. In Fig. 9.16, a membrane initially at a potential of -60 mV is voltage clamped to a new value of 0 mV (pulse 1). With depolarization, there is a rapid increase in Na$^+$ permeability, followed by its spontaneous decay. When this first pulse is followed by an identical pulse (pulse 2) to the same level of membrane potential soon thereafter (Fig. 9.16B), there is still an increase in Na$^+$ permeability, but the increase is much smaller than it was for the first stimulus. Indeed, when the separation between these pulses is reduced further, a point is reached where there is absolutely no change in Na$^+$ permeability produced by the second depolarization (Fig. 9.16C). The two pulses must be separated by several milliseconds before the change in Na$^+$ permeability is equal to that obtained initially (Fig. 9.16A). How do we explain these results, and what do they have to do with the absolute refractory period? Just as it takes a certain amount of time for the Na$^+$ channels to inactivate, it also takes some time for these channels to recover from the inactivation and be able to respond again to a second depolarization. Therefore, as a result of initiating one action potential, there is an increase in Na$^+$

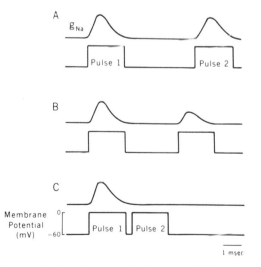

FIG. 9.16. Recovery from inactivation. The second of two depolarizing pulses activates less Na$^+$ conductance as the interval between the pulses decreases. Once the Na$^+$ channels become inactivated by the first pulse, several milliseconds are required before they recover completely. (See text for explanation.)

permeability that spontaneously inactivates. As a result, if one attempts to initiate a second action potential soon after the first, the Na^+ permeability will not have recovered from inactivation, making a second depolarization ineffective in initiating a voltage-dependent change in Na^+ permeability. If there is no voltage-dependent change in Na^+ permeability, no action potential will be produced. Thus, the absolute refractory period is most simply understood in terms of this process of recovery from Na^+ inactivation. The recovery from Na^+ inactivation may contribute to the relative refractory period as well. During this recovery time, the threshold for an action potential will be higher because greater depolarization will be required to activate sufficient Na^+ influx to exceed the K^+ efflux.

Accommodation

Accommodation is defined as a change in the threshold of an excitable membrane when slow depolarization is applied. In the previous examples, rapid depolarization is applied, and the threshold occurs at a relatively fixed membrane potential (e.g., Fig. 8.3). When a slowly developing depolarization is applied, however, the threshold is frequently at a more depolarized level, and indeed, if the depolarization is slow enough, no action potential will be initiated despite the level of depolarization. The process of Na^+ inactivation contributes to the phenomenon of accommodation. Essentially, slow depolarization provides sufficient time for the Na^+ channels to inactivate before they can be sufficiently activated. In terms of the molecular model of Fig. 9.10, there are an insufficient number of Na^+ channels in state B because they are already in state C.

BIBLIOGRAPHY

Aidley DJ. *The physiology of excitable cells*, 2nd ed. Cambridge; Cambridge University Press, 1978; Chapter 5.
Hille B. *Ionic channels of excitable membranes*, 2nd ed. Sunderland, MA; Sinauer Associates, 1992; Chapters 2 and 3.
Kandel ER, Schwartz JH, Jessell TM. *Principles of neural science*, 3rd ed. New York; Elsevier/North-Holland, 1991; Chapter 8.
Katz B. *Nerve muscle and synapse*. New York; McGraw-Hill, 1966; Chapters 4 and 5.
Nicholls JG, Martin AR, Wallace BG. *From neuron to brain*, 3rd ed. Sunderland, MA; Sinauer Associates, 1992; Chapter 4.
Schmidt RF, ed. *Fundamentals of neurophysiology*, 2nd ed. New York; Springer-Verlag, 1978; Chapter 2.

ADDITIONAL READING

Adams DJ, Smith SJ, Thompson SH. Ionic currents in molluscan soma. *Ann Rev Neurosci* 1980;3:141.
Atkinson NS, Robertson GA, Ganetzky B. A component of calcium-activated potassium channels encoded by the *Drosophila slo* locus. *Science* 1991;253:551–555.

Bernstein J. Untersuchungen zur Thermodynamik der bioelektrischen Strome. Erster Theil. *Pflug Arch Ges Physiol* 1902;92:521.

Caterrall WA. Structure and function of voltage-sensitive ion channels. *Science* 1988;242:50–61.

Hodgkin AL, Huxley AF. Currents carried by sodium and potassium ions through the membrane of the giant axon of *Loligo*. *J Physiol* 1952;116:449.

Hodgkin AL, Huxley AF. The components of membrane conductance in the giant axon of *Loligo*. *J Physiol* 1952;116:473.

Hodgkin AL, Huxley AF. The dual effect of membrane potential on sodium conductance in the giant axon of *Loligo*. *J Physiol* 1952;116:497.

Hodgkin AL, Huxley AF. A quantitative description of membrane current and its application to conduction and excitation in nerve. *J Physiol* 1952;117:500.

Hodgkin AL, Katz B. The effect of sodium ions on the electrical activity of the giant axon of the squid. *J Physiol* 1949;108:37.

Keynes RD. The ionic movements during nervous activity. *J Physiol* 1951;114:119.

Li M, West JW, Numann R, Murphy BJ, Scheuer T, Catterall WA. Convergent regulation of sodium channels by protein kinase C and cAMP-dependent protein kinase. *Science* 1993;261:1439–1442.

Nastuk WL, Hodgkin AL. The electrical activity of single muscle fibres. *J Cell Comp Physiol* 1950; 35:39.

Overton E. Beitrage zur allgemeinen Muskel-und Nervenphysiologie. *Pflug Arch Ges Physiol* 1902; 92:346.

Rudy B, Kentros C, Vega-Saenz de Miera E. Families of potassium channel genes in mammals: toward an understanding of the molecular basis of potassium channel diversity. *Mol Cell. Neurosci* 1991; 2:89–102.

Sigworth FJ, Neher E. Single Na^+ channel currents observed in cultured rat muscle cells. *Nature* 1980; 287:447–449.

Snutch TP, Reiner PB. Ca^{2+} channels: diversity of form and function. *Curr Opin Neurobiol* 1992; 2:247–253.

Stühmer W, Parekh AB. The structure and function of Na^+ channels. *Curr Opin Neurobiol* 1992;2: 243–246.

Tsien RW, Ellinor PT, Horne WA. Molecular diversity of voltage-dependent Ca^{2+} channels. *TIPS* 1991;12:349–354.

West JW, Patton DE, Scheuer T, Wang Y, Goldin AL, Catterall WA. A cluster of hydrophobic amino acid residues required for fast Na^+-channel inactivation. *Proc Natl Acad Sci USA* 1992;89: 10910–10914.

10

Propagation of the Action Potentials

Up to this point, we have been considering the ionic mechanisms that underlie the action potential in a dimensionless nerve cell. One of the interesting features of action potentials, however, is that not only are they elicited in an all-or-nothing fashion, but they also are propagated in an all-or-nothing fashion. If an action potential is initiated in the cell body, it will propagate without decrement along the axon and eventually invade the synaptic terminal. An action potential recorded in the nerve axon has an amplitude and time course identical to the action potential that was initiated in the cell body. It is the ability of action potentials to propagate in this all-or-nothing fashion that endows the nervous system with the capability to transmit information over long distances.

BASIC PRINCIPLES

To begin to consider the mechanisms that account for the propagation of the action potential, it is useful to examine the charge distribution that is found in an isolated section of a nerve axon. When an axon is at rest, the potential of the inside is negative with respect to the outside (Fig. 10.1A). The distribution of charge is simply due to the tendency of K^+ to diffuse from its region of high concentration inside the axon to its region of low concentration outside the axon. Consider what will happen if at some point along the axon an action potential is initiated. At the peak of the action potential the inside of the cell will be positive with respect to the outside (Fig. 10.1B). At this point there is a new charge distribution at a localized portion of the membrane. Adjacent regions of the axon, however, still have their initial charge distribution (inside negative). Unlike charges attract each other, so the positive charge produced by the action potential will tend to move toward the adjacent region of membrane (still at rest), which has a negative charge (Fig. 10.1C). As a result of this positive charge movement the adjacent region of the axon will become depolarized. If sufficient charge moves to depolarize the adjacent portion of the membrane to threshold and elicit voltage-dependent changes in Na^+ per-

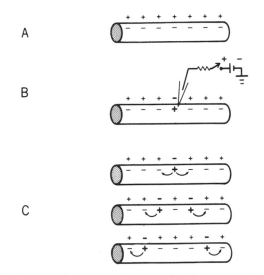

FIG. 10.1. Schematic diagram of sequence of steps underlying propagation of the action potential. **(A)** Rest. **(B)** Initiation. **(C)** Propagation.

meability, a new action potential at this adjacent region will be initiated. So as a result of an action potential in one portion of an axon and the subsequent charge transfer along the surface of the membrane, a "new" action potential will be generated. This new action potential then will cause charge transfer to its adjacent region causing, in a sense, another new action potential to be initiated. It should be clear that this process once initiated will propagate all the way to the end of the axon.

DETERMINANTS OF PROPAGATION VELOCITY

What are the factors that determine the rate of propagation of the action potential? To address this question, some details of the passive properties of axonal membranes must be examined. Two important passive properties that are directly related to the rate at which an axon can propagate action potentials are the space (or length) constant and the time constant. The space and time constants are known as passive properties because they are not *directly* dependent on metabolism or any voltage-dependent permeability changes such as those that underlie the action potential. These are intrinsic properties that are reflections of the physical properties of the neuronal membrane. Indeed, they are properties of all membranes.

Time Constant

To consider the time constant, first consider a very simple thermal analog. Take the case where a block of metal that is initially at 25°C is placed on a hot plate that is at 50°C. Assume that the hot-plate temperature is constant. What will be the consequences of placing the block on the hot plate? It is obvious that over a period of time, the temperature of the block will change from its initial value of 25°C to a final value of 50°C, but it will not do so instantaneously. It will take a certain period of time for the heat transfer to occur. If the temperature of the block is measured as a function of time, the temperature will change as an exponential function of time, approaching a final value of 50°C.

A similar phenomenon is observed in membranes when one applies an artificial depolarizing or hyperpolarizing stimulus. This is illustrated in Fig. 10.2. A nerve cell is impaled with one electrode to record the membrane potential and another electrode to depolarize or hyperpolarize the cell artificially (Fig. 10.2A). The cell is initially at its resting potential of − 60 mV. At time zero, the stimulating electrode is connected to a battery. The size of the battery is such that the stimulus will eventually depolarize the cell by 10 mV (Fig. 10.2B). As a result, the membrane potential will change from its initial value of − 60 mV to a final value of − 50 mV. Note that even though the current flow is instantaneous (and constant), the membrane potential does not change instantaneously. There is a period of time during which the membrane potential charges to its new final level of − 50 mV. This charging process is an exponential function of time. Just as the temperature in the block changed exponentially as a function of time, the change in potential in the neuron produced by an applied depolarization (or hyperpolarization) also follows an exponential time course. For such exponential functions of time it is possible to

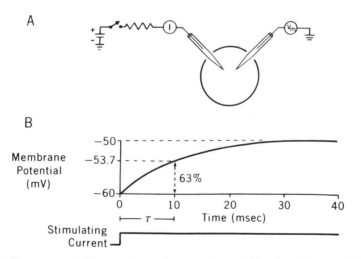

FIG. 10.2. Measurement of the membrane time constant. **(A)** Experimental setup. **(B)** Change in membrane potential as a function of time after the delivery of a constant step of depolarizing current. (Modified from E. R. Kandel, *The cellular basis of behavior.* San Francisco; Freeman, 1976: Chapters 5 and 6.)

define what is known as a time constant. The time constant refers to the time it takes for the potential change to reach 63% of its final value.

The general equation for a response that changes as an exponential function of time is

$$\% \text{ Response} = 100(1 - e^{-t/\tau}) \qquad [10.1]$$

where t is time and τ is the time constant. At time 0, e^{-0} is equal to 1, and the % response is zero. At infinite time, $e^{-\infty}$ is equal to zero, so the % response is 100. A special case occurs when $t = \tau$. Then, $e^{-1} = 0.37$, and the % response $= 63\%$. A detailed understanding of the mathematics is not important. What is important is that the time constant is a simple index of how rapidly a membrane will respond to a stimulus.

The time constant for the cell illustrated in Fig. 10.2B is 10 msec. Thus, in 10 msec the potential has changed from -60 mV to -53.7 mV (63% of its final displacement). The smaller the time constant of a cell, the more rapidly the cell will respond to an applied stimulus. If the time constant were 1 msec, the potential change would occur very rapidly, and the cell would reach a value of -53.7 mV in just 1 msec. If the time constant were 40 msec, the potential change would occur very slowly, and the potential would reach -53.7 mV in 40 msec.

A simple formula that describes the time constant in terms of the physical properties of the membrane is

$$\tau = R_m C_m \qquad [10.2]$$

Here, R_m simply reflects the resistive properties of the membrane and is equivalent to the inverse of the permeability, since the less permeable the membrane the higher the resistance. C_m represents the membrane capacitance. This is a physical parameter that describes the ability of a membrane to store charge. It is equivalent to the ability of the metal block to store heat. The larger the size of a block, the better able it is to store heat. Similarly, the larger the membrane capacitance, the better able it is to store charge.

Space Constant

Before discussing how the time constant is related to propagation velocity, the other passive membrane property, the space (or length) constant, will be discussed. To introduce this phenomenon, it is useful to turn again to a thermal analog. Instead of considering a small block on a hot plate, consider what might happen when one end of a long metal rod touches the hot plate. The hot plate is at 50°C, and the rod is initially at 25°C. If the rod is placed in contact with the hot plate and a sufficient period of time elapses for the temperature changes to stabilize, what will be the temperature gradient along the rod? It is obvious that the temperature at the end of the rod in contact with the hot plate will be 50°C (the same temperature as the hot plate). The temperature of the rod, however, will not be 50°C along its length. The

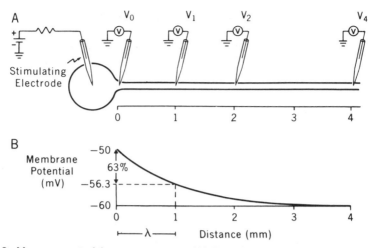

FIG. 10.3. Measurement of the space constant. **(A)** Experimental setup. **(B)** Changes in membrane potential as a function of distance along the axon. A constant depolarizing current is applied to the cell body to depolarize the cell body from rest (-60 mV) to -50 mV. (Modified from E. R. Kandel, *The cellular basis of behavior.* San Francisco; Freeman, 1976: Chapters 5 and 6.)

temperature near the hot plate will be 50°C, but along the rod the temperature will gradually fall, and if the rod is long enough the temperature may still be 25°C at its end. If the temperature of the rod at various distances from the hot plate is measured, the temperature will be found to decay as an exponential function of distance.

Just as there is a spatial degradation of temperature in a long rod, there is also a spatial degradation of potential along a nerve axon, which is referred to as electrotonic conduction. Figure 10.3A illustrates how it is possible to demonstrate this. One electrode is in the cell body and will be used to depolarize the cell artificially. A number of other electrodes are placed at various distances along the axon to record the potential gradient as a function of distance from the cell body. Initially, the cell body and all regions of its axon are at the resting potential of -60 mV. A sufficient subthreshold depolarization is then applied to the cell body to depolarize the cell body to -50 mV. Just as one end of the rod was placed on a 50°C heat source, the cell body is forced to a potential of -50 mV that is different from its resting level. After waiting a sufficient period of time for the potential changes to stabilize, the measurements are made. Very near the cell body the potential is -50 mV (Fig. 10.3B). Because the membrane potential is sampled at points away from the cell body, however, there is a change in the electronic potential from its value or -50 mV in the cell body to more negative values. Measurements made a great enough distance from the cell body reveal that the potential recorded is the resting potential (-60 mV). The potential profile is an exponential function of distance and a space constant (denoted by the symbol λ) can be defined. The space constant is the distance it takes for the depolarizing displacement (i.e., 10 mV) to decay by 63% of its initial value. In this particular cell the space constant is 1 mm. This means that 1 mm away from the cell body the potential would have changed from its value of -50 mV in the cell body to a value of -56.3 mV in the axon. The greater the space constant, the greater will be the extent of the propagation of this electrotonic potential. If the space constant were 2 mm, this potential profile would decay less so that at 2 mm the potential would be at -56.3 mV.

Just as it is possible to provide a formula for the time constant in terms of the physical properties of the membrane, it is also possible to derive a formula for the space constant. The space constant is equal to

$$\lambda = \sqrt{\frac{dR_m}{4R_i}} \qquad [10.3]$$

where R_m once again refers to the resistive properties of the membrane (the inverse of the membrane permeability); R_i is a term that refers to the resistive properties of the intracellular medium (the resistance of the axoplasm to the flow of ions); and d is the diameter of the axon.

Relationship Between Propagation Velocity and the Time and Space Constants

It is possible to make some qualitative predictions about the way in which the space and time constants affect propagation velocity. If the space constant is large, a potential produced at one portion of an axon will spread greater distances along the axon. Since the potential will spread a greater distance along the axon, it will bring distant regions to threshold sooner. Thus, the greater the space constant, the greater will be the propagation velocity. The time constant is a reflection of the rate that a membrane can respond to an applied stimulus current. The smaller the time constant, the greater will be the ability of a membrane to respond rapidly to stimulus currents. Action potentials will be initiated sooner, and the propagation velocity will be greater. Therefore, the smaller the time constant, the greater will be the propagation velocity.

Thus, the propagation velocity is directly proportional to the space constant but inversely proportional to the time constant. Since relationships for both the space and time constants are known, it is possible to derive a new formula that describes the propagation velocity:

$$\text{Velocity} \propto \frac{\sqrt{dR_m/4R_i}}{R_mC_m} \qquad [10.4]$$

Thus,

$$\text{Velocity} \propto \frac{1}{C_m}\sqrt{\frac{d}{4R_mR_i}} \qquad [10.5]$$

It is desirable to have axons that have high propagation velocities, since there is great survival value to rapid information transmission. For example, to initiate a motor response to some noxious stimulus, such as touching a hot stove, action potentials must propagate rapidly along sensory and motor axons.

Given that the propagation velocity can be described in terms of the physical properties of nerve axons, we can begin to examine strategies utilized by the nervous system to endow axons with high propagation velocities. One of the simplest and most obvious ways of doing this is to increase the diameter of the axon. By increasing the diameter, the propagation velocity is increased. This is exactly the strategy that has been used extensively by many invertebrate axons, of which the squid giant axon is the prime example. The giant squid axon has a diameter of about 1 mm, which endows it with perhaps the highest propagation velocity of any invertebrate axon. There is one severe price that is paid, however, when the propagation velocity is increased in this way. The key to understanding this problem is the square root relationship in the formula for propagation velocity. The square root relationship requires that to double the propagation velocity the fiber diameter must be quadrupled.

Therefore, to get moderate increases in propagation velocity one has to increase axons to very large diameters. Although this is frequently observed in invertebrates, it is not generally utilized in the vertebrate central nervous system. For example, it is known that the propagation velocity of axons in the optic tract is about the same as that of the squid giant axon. If all the axons in the optic tract were the size of the squid giant axon, however, the optic tract by itself would take up the space of the entire brain.

Conduction in Myelinated Axons

Clearly, there must be another means available by which axons can increase their propagation velocity without drastic changes in fiber diameter. You will note from the relationship for propagation velocity that by changing the membrane capacitance, velocity can be affected directly without involving the square-root relationship. It is possible to decrease the membrane capacitance simply by coating the axonal membrane with a thick insulating sheath. This is exactly the strategy used by the vertebrates. Many vertebrate axons are coated with a thick lipid layer known as myelin. As a result of myelin, the capacitance is greatly reduced, and propagation velocity is greatly increased. In principle, there is one severe problem with increasing the propagation velocity with this technique. Coating the axon with a lipid layer would tend to cover the channels or pores in the membrane that endow the axon with the ability to initiate and propagate action potentials. The nervous system has solved this problem by only coating portions of the axon with myelin. Certain regions called nodes are not covered. At these bare regions, voltage-dependent changes in membrane permeability take place that generate action potentials.

The process of conduction in myelinated fibers is illustrated in Fig. 10.4. The

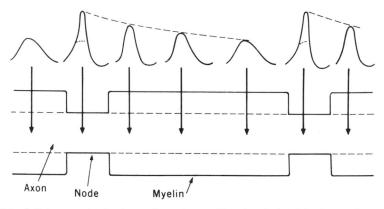

FIG. 10.4. Saltatory conduction in myelinated axon. The electrical activity "jumps" from node to node. Between the nodes the action potential is propagated electrotonically with little decrement. At the nodes, where ionic current can flow, the electronic potential reaches threshold and triggers an action potential. The process is then repeated. *Arrows* indicate regions of the axon where the potentials were recorded.

dashed lines show a nerve axon that is covered with a layer of myelin. Note that the myelin does not cover the entire axon; there are bare regions or nodes where voltage-dependent changes in permeability can take place and action potentials can be elicited. Assume there is an action potential elicited at the node to the left. As a result of the action potential, there is a large depolarization. The inside of the cell becomes positive with respect to the outside. The action potential cannot propagate along the myelinated region via the active process that was described earlier, simply because the voltage-dependent changes in permeability cannot take place; however, the action potential can conduct passively. That conduction will occur very rapidly because the membrane capacitance is reduced. Because of the small amount of decrement, the potential that emerges at the next node will be of a sufficient level to depolarize the next node to threshold. A new action potential will be initiated, and the process will be repeated.

The type of propagation that occurs in myelinated fibers is known as *saltatory conduction* because the action potential appears to "jump" from node to node. At the nodes, there are voltage-dependent changes in membrane permeability, whereas in the internodal regions, the potential is conducted in a passive fashion. No voltage-dependent changes in permeability take place in the internodal region.

BIBLIOGRAPHY

Aidley DJ. *The physiology of excitable cells*, 2nd ed. Cambridge: Cambridge University Press, 1978; Chapter 4.
Hodgkin AL. *The conduction of the nervous impulse*. Springfield: Charles C Thomas, 1984.
Jack JJB, Nobel D, Tsien RW. *Electric current flow in excitable cells*. Oxford: Clarendon Press, 1975.

Shepherd GM, ed. *The synaptic organization of the brain*, 3rd ed. New York: Oxford University Press, 1990: Chapter 13.

Nicholls JG, Martin AR, Wallace BG. *From neuron to brain*, 3rd ed. Sunderland, MA: Sinauer Associates, 1992: Chapter 5.

ADDITIONAL READING

Huxley AF, Stampfli R. Evidence for saltatory conduction in peripheral myelinated nerve-fibres. *J Physiol* 1949;108:315.

Huxley AF, Stampfli R. Saltatory transmission of the nervous impulse. *Arch Sci Physiol* 1949;3:435.

Rushton WAH. A theory of the effects of fibre size in medullated nerve. *J Physiol* 1951;115:101.

Tasaki I. Conduction of the nerve impulse. *Handbook of physiology. Section 1: Neurophysiology*, vol 1. Baltimore: Williams & Wilkins, 1959:75.

11

Extracellular Recordings from Excitable Membranes

In many cases, it is not possible to monitor the electrical activity of nerve cells using intracellular microelectrode recording techniques; however, it is possible to measure a reflection of that intracellular activity from the immediate extracellular environment of excitable cells. Typical examples of this type of recording are the electromyogram (EMG), the electrocardiogram (EKG), electro-oculogram (EOG), and electroencephalogram (EEG). Each of these recordings reflects the intracellular activity from a host of simultaneously excited cells.

EXTRACELLULAR RECORDING FROM A SINGLE AXON

To introduce the principles of extracellular recording, it is useful to begin with the simplest case—the extracellular recording of activity from a single nerve axon. Figure 11.1 illustrates both the general strategy and typical results. Two metal electrodes, A and B, are placed in close proximity to an isolated nerve axon. They do not impale the axon but are very near its outside surface. The electrodes are connected to a suitable electronic amplifier that is designed so that the voltage displayed is the difference between the voltage sensed at electrode B (V_B) and the voltage sensed at electrode A (V_A):

$$V = V_B - V_A \qquad [11.1]$$

The recording device measures voltages B and A and subtracts the two to generate a visual display. In this example, the display is such that an upward deflection on the recording device reflects a positive difference (e.g., $V_B - V_A = +$), and a downward deflection reflects a negative difference (e.g., $V_B - V_A = -$).

Let us first examine the case where there is no action potential propagating along the nerve axon (Fig. 11.1, trace 1). At rest, the membrane potential of the axon is approximately -60 mV inside with respect to the outside. Although this is not the usual convention, another way of stating it is that the axon is $+60$ mV outside with

113

respect to the inside. Thus, the electrodes placed in the immediate environment of the cell membrane will both sense equal positive charge on the outside of the membrane. By taking the difference between the potentials at the recording electrodes, the electronic amplifier will record a potential difference of 0. This will correspond to no deflection on the recording device. To summarize, at rest,

$$V = V_B - V_A$$

at point 1 (Fig. 11.1) and

$$V = (+) - (+) = 0 = \text{no deflection}$$

Let us now initiate an action potential at some distant point to the left of the axon illustrated in Fig. 11.1 and allow that action potential to propagate along the nerve axon toward the recording electrodes. Consider the consequences of the changes in charge distribution when the wave of depolarization representing the propagating action potential first comes near the region of electrode A. Assume for the moment that we can "freeze" the action potential at this point in time. We have learned that during the peak of the action potential, the potential inside the cell becomes approximately $+55$ mV with respect to the outside. Stated in a different way, the outside of the axon will be negative with respect to the inside. So, electrode A will now sense a negative outside surface charge, but electrode B will sense the same positive outside charge (since the action potential has not yet propagated to electrode B). Thus, at point 2 (Fig. 11.1), the electronic amplifier will take the difference between the two potentials and determine a positive difference. The positive difference will correspond to an upward deflection of the recording device. To summarize, at point 2 (Fig. 11.1),

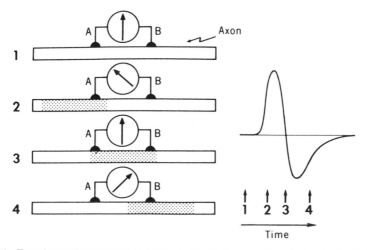

FIG. 11.1. Experimental arrangement for the extracellular recording of electrical activity from a single nerve axon (see text for explanation).

$$V = V_B - V_A$$

$$V = (+) - (-) = + = \text{upward deflection}$$

If we now "unfreeze" the action potential and allow the wave of depolarization to continue to propagate, it eventually will reach a point where the action potential is in the region of both electrodes A and B (trace 3, Fig. 11.1). We now "freeze" the action potential at the point where each electrode senses equal parts of the action potential. The regions under both electrodes will be positive inside with respect to the outside or negative outside with respect to the inside. So, electrode B will sense a negative charge, and electrode A will sense the same negative charge. The electronic amplifier will subtract the two and determine that there is no difference. The recording device will return to its initial state and display no deflection. To summarize, at point 3 (Fig. 11.1),

$$V = V_B - V_A$$

$$V = (-) - (-) = 0 = \text{no deflection}$$

The potential recorded at point 3 is an interesting case. The recording device is back to its initial state, but that does not mean that the recorded potential is equal to the resting potential. Rather, it means that electrodes A and B are recording the same potential.

What happens as the action potential continues to propagate along the axon? The area of excitation under the two electrodes will move away from electrode A and will eventually reach a point where the excited region (the action potential) is predominantly in the vicinity of electrode B. As a result, the region under electrode A will repolarize so that the inside of the axon will become negative with respect to the outside, or the outside will be positive with respect to the inside (as it was initially). The region under electrode B will still be excited so that it will be negatively charged. Thus, at point 4 (Fig. 11.1), the difference between the potentials at the two electrodes is negative, and a negative difference corresponds to a downward deflection of the recording device: At point 4,

$$V = V_B - V_A$$

$$V = (-) - (+) = - = \text{downward deflection}$$

Eventually, the excited region will move away from the electrodes, and both electrodes will sense positive charge. The difference between the two will be zero, and the recording device will return to its initial state.

Thus, by using extracellular recording techniques, it is possible to obtain signals that correspond to the passage of an action potential along a nerve axon. Note that the form of the measured potential is very different from the action potential recorded with an intracellular microelectrode. In addition, it is only a rough reflection of the magnitude and duration of the underlying membrane permeability changes. This type of recording is used primarily as a simple index of whether or not an action potential has occurred.

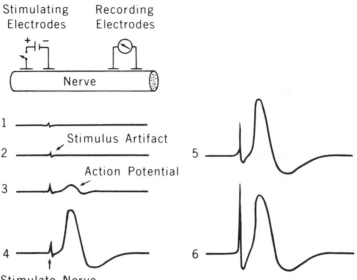

FIG. 11.2. Extracellular recording of the electrical activity from a nerve. One pair of electrodes is placed so that the nerve can be electrically stimulated while a second pair of electrodes records the neural activity. With weak intensity stimuli (*traces 1 and 2*), only stimulus artifacts are produced, and no neural activity is observed. As the stimulus intensity to the nerve is increased, the amplitude of the action potential increases (*traces 3–5*) until a point is reached where further increases in intensity (although producing a larger artifact) do not produce a larger action potential (*traces 5 and 6*). (Modified from B. Katz, *Nerve muscle and synapse.* New York: McGraw-Hill; 1966: Chapter 2.)

EXTRACELLULAR RECORDING FROM A NERVE BUNDLE

Figure 11.2 illustrates stimulation and recording of a nerve bundle rather than a single axon. Recall that a nerve bundle, or nerve, is made up of a group of individual axons.[1] Stimulating electrodes are placed near one end of the nerve bundle to depolarize the axons to threshold and initiate action potentials, and recording electrodes are placed at the other end to record the resulting extracellular potential changes produced by the action potentials propagating along the nerve. Trace 1 of Fig. 11.2 shows that, when a very weak stimulus is delivered, no action potential is recorded. There is only a small transient diphasic deflection that is due to interference picked up by the recording electrodes from the stimulating electrodes. The small initial deflection is known as a stimulus artifact. Increasing the stimulus intensity further produces a larger artifact, but still no action potential is produced (trace 2). However, as the intensity is increased further, a point is reached where a di-

---

[1]Note the distinction between a nerve and an axon. The term *nerve* generally refers to a nerve bundle (e.g., the vagus nerve), which contains many individual axons.

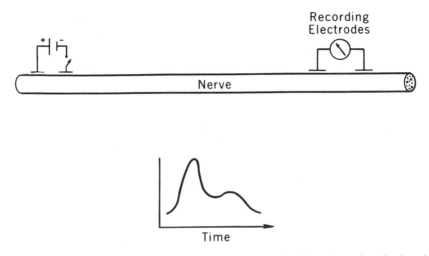

FIG. 11.3. Recording the compound action potential at a point distant from the stimulus site. Multiple peaks can be observed due to the contributions of fibers with different conduction velocities.

phasic action potential is recorded (trace 3, Fig. 11.2). Note that there is a delay between the stimulus artifact and the diphasic action potential, which is a reflection of the time it takes for the action potential to propagate from its site of initiation (at the stimulating electrodes) to the recording electrodes.[2] An interesting observation occurs when the stimulus intensity is increased further. Now the amplitude of the extracellular action potential is increased (trace 4, Fig. 11.2). Increasing the stimulus intensity further produces yet a greater increase in the size of the action potential (trace 5, Fig. 11.2). Eventually, a point is reached where increasing the stimulus intensity produces no further increase in the size of the action potential.

At first these results may seem somewhat paradoxical because they appear to contradict the all-or-nothing law of the action potential. Recall that an action potential is not only propagated in an all-or-nothing fashion but is also initiated in an all-or-nothing fashion. We learned earlier that an increase in stimulus intensity beyond threshold produces an action potential identical in its amplitude and time course to the action potential produced with a threshold stimulus. In Figure 11.2, the action potential amplitude increases as a function of the stimulus intensity. The key to understanding this apparent paradox is that we are now dealing with a nerve bundle that contains many different axons. The individual axons in the nerve have different diameters and therefore different thresholds to extracellularly applied stimulating currents. As a result, when the stimulus intensity is increased, the first axon to initiate an action potential is the nerve axon that has the lowest threshold. That

---

[2]Knowing the time delay and the distance between the stimulating and recording electrodes, the propagation velocity of the action potentials can be calculated.

action potential is then sensed by the recording electrodes and displayed on the recording device. As the stimulus intensity is increased further, more axons are brought to threshold so that there are multiple action potentials propagating along the nerve bundle, and these individual action potentials will each make a contribution to the signal sensed by the recording electrodes. Eventually, a stimulus intensity is reached that brings all the axons above threshold so that further increases in the stimulus intensity initiate no additional action potentials, and the size of the extracellularly recorded action potential remains at its peak value. The action potential that one records with extracellular electrodes from a nerve bundle is known as a *compound action potential* (compound because it is a summation of contributions from the individual action potentials in the axons that compose the nerve bundle).

One additional variant on this theme is illustrated in Fig. 11.3. This is a case where the stimulating and recording electrodes are separated by a considerable distance. An interesting observation is that the potential change recorded can be quite different from the shape of the potential changes seen in Figs. 11.1 and 11.2. Previously, there was a relatively smooth rise, fall, and then recovery—a diphasic action potential. In this new case, the action potential has multiple peaks (in some cases, there can be three or more different peaks). What is the origin of these different peaks? The important point is that a nerve bundle may contain many different axons, many of which have different diameters. In addition, some of these axons may be myelinated and others unmyelinated. As a result, action potentials that are initiated in these axons will have different propagation velocities. The action potentials will reach the recording electrodes at different times so the different peaks reflect differences in the time of arrival of action potentials. The situation is somewhat analogous to a fast runner and a slow runner in a race. Initially, they are both at the starting line, but over a period of time the faster runner outdistances the slower runner and reaches the finish line sooner. The stimulating electrodes are analogous to the starting line, whereas the recording electrodes are analogous to the finish line.

BIBLIOGRAPHY

Katz B. *Nerve muscle and synapse*. New York: McGraw-Hill, 1966; Chapter 2.
Ruch T, Patton HD, eds. *Physiology and biophysics*. Philadelphia: WB Saunders Co, 1982; Chapter 4.

12

Neuromuscular and Synaptic Transmission

Nerve cells are capable not only of initiating action potentials but also of communicating directly with other cells. Transferring information from one nerve cell to another or from one nerve cell to a muscle cell is a process known as synaptic transmission. Synaptic transmission is mediated by specialized junctions called synapses.

TYPES OF SYNAPTIC TRANSMISSION

There are two distinct categories of synaptic transmission: chemical and electrical. Each category is associated with a number of characteristic morphological and physiological properties. Features of these two types of synaptic transmission are schematically illustrated in Fig. 12.1. Electrical synaptic transmission is mediated by specialized structures known as gap junctions. The gap junctions associated with electrical synapses provide a pathway for cytoplasmic continuity between the presynaptic and the postsynaptic cells. As a result, a depolarization (or a hyperpolarization) produced in the presynaptic terminal produces a change in potential of the postsynaptic terminal. The potential change in the postsynaptic cell is due to the direct ionic pathway between the presynaptic and postsynaptic cell. For electrical synapses there is a minimal synaptic delay. As soon as a potential change is produced in the presynaptic terminal, there is a reflection of that potential change in the postsynaptic cell. Electrical junctions are found not only in the nervous system but also between other excitable membranes, such as smooth muscle and cardiac muscle cells. In these muscle cells, they provide an important pathway for the propagation of action potentials from one muscle cell to another.

For chemical synapses, there is a distinct cytoplasmic discontinuity that separates the presynaptic and postsynaptic membranes. This discontinuity is known as the synaptic cleft. The presynaptic terminal of chemical synapses contains a high concentration of mitochondria and synaptic vesicles, and there is a characteristic thickening of the postsynaptic membrane. As a result of a depolarization or an action

A

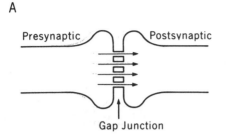

B

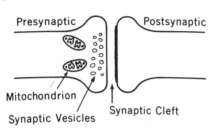

FIG. 12.1. Schematic diagram of two types of synaptic junctions. **(A)** Electrical synapse. Transmission takes place through gap junctions that provide an ionic pathway between the pre- and postsynaptic junctions. **(B)** Chemical synapse. Neurotransmitter is released from the presynaptic terminal, which diffuses across the synaptic cleft and interacts with receptor sites on the postsynaptic membrane.

potential in the presynaptic terminal, chemical transmitters are released from the presynaptic terminal, which diffuse across the synaptic cleft and bind to receptor sites on the postsynaptic membrane. This leads to a permeability change that produces the postsynaptic potential. For chemical synapses, there is a delay (usually, approximately 0.5–1 msec in duration) between the initiation of an action potential in the presynaptic terminal and a potential change in the postsynaptic cell. The synaptic delay is due to the time necessary for transmitter to be released, diffuse across the cleft, and bind with receptors on the postsynaptic membrane. Chemical synaptic transmission is generally unidirectional. A potential change in the presynaptic cell releases transmitter that produces a postsynaptic potential, but a depolarization in the postsynaptic cell does not produce any effects in the presynaptic cell because no transmitter is released from the postsynaptic cell at the synaptic region.

In the early part of this century, most physiologists believed that all synaptic transmission in the nervous system occurred via electrical synapses. In the 1930s and 1940s, chemical transmitter substances began to be discovered, and the pendulum swung in the opposite direction, so that most physiologists believed that synaptic transmission occurred exclusively via chemical synapses. It is now known that there are both electrical and chemical synapses in the nervous system. By far, the most predominant type of synapse is the chemical synapse, and for that reason this and the following two chapters focus on chemical synaptic transmission.

SYNAPTIC TRANSMISSION AT THE
NEUROMUSCULAR JUNCTION

The chemical synapse from which most of our information about synaptic transmission has been derived is the synaptic contact made by a spinal motor neuron with a skeletal muscle cell. Figure 12.2 is a schematic diagram of some of the general features of this synapse. The myelinated axon of a motor neuron whose cell body is located in the ventral horn of the spinal cord innervates a number of individual muscle fibers (Fig. 12.2A). At each muscle fiber the axon branches further and forms a series of contacts with the muscle cell. An expanded view of one such synaptic contact, which illustrates some of the characteristic morphological features of the chemical synapse, is shown in Fig. 12.2B. There are (1) a large concentration of small vesicles and mitochondria in the presynaptic terminal, and (2) a distinct synaptic cleft that separates the presynaptic terminal from the postsynaptic cell (in this case, the skeletal muscle cell). This particular synapse, because of its very characteristic shape, is known as the motor end plate. It is also known as the neuromuscular junction because it is the junction made by a motor axon with a muscle cell. Additional morphological features of the synapse at the neuromuscular junction are illustrated by the expanded view in Fig. 12.2C. Here it is seen that (1) the

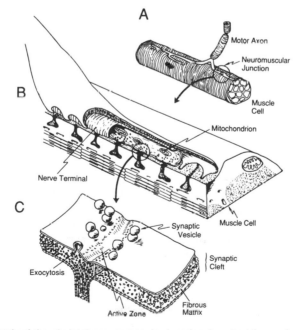

FIG. 12.2. Sketch of the skeletal neuromuscular junction (see text for explanation). (Modified from H. A. Lester. *Sci Am* 1977;236:106; and B. Hille, *Ionic Channels of excitable membranes.* Sunderland, MA; Sinauer, 1984: Chapter 6.)

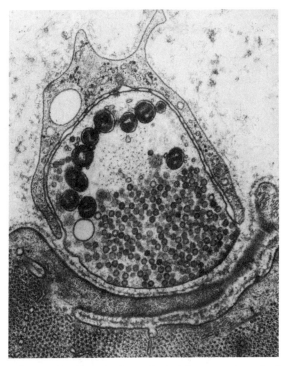

FIG. 12.3. Electron micrograph of a synaptic contact at the neuromuscular junction. The presynaptic terminal (*upper portion*) contains many small vesicles and larger mitochondria. There is a distinct separation (the synaptic cleft) between the membranes of the pre- and postsynaptic cells. (Micrograph produced by Dr. John E. Heuser of Washington University School of Medicine, St. Louis, MO.)

synaptic vesicles are clustered in distinct regions of the terminal known as the active zone; (2) the active zone is opposite an invagination of the muscle membrane known as the junctional fold; and (3) there is a characteristic thickening of the postsynaptic membrane, perhaps due to the high density of postsynaptic receptors for acetycholine (ACh). In addition, the synaptic cleft is filled with a fibrous matrix and the enzyme acetylcholinesterase (AChE). The functions of ACh and AChE are described below. Figure 12.3 is an electron micrograph of the synaptic junction illustrating the high concentration of vesicles and mitochondria in the presynaptic terminal, the fibrous matrix in the synaptic cleft, and the characteristic thickening of the postjunctional membrane.

THE END-PLATE POTENTIAL

It is possible to study various aspects of chemical synaptic transmission in a reduced preparation. One can dissect a skeletal muscle cell (with its intact neural

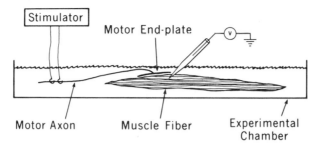

FIG. 12.4. Schematic diagram of the preparation used to study features of synaptic transmission at the skeletal neuromuscular junction.

innervation) from an animal and place it in an experimental solution, where it can remain viable for long periods of time (Fig. 12.4). The postsynaptic muscle cell is impaled with a microelectrode to record potentials in the muscle cell, and the motor axon is stimulated to initiate action potentials in the axon. Figure 12.5A illustrates a typical result.

At the arrow, an electrical stimulus is delivered to the motor axon. This stimulus elicits an action potential in the axon, which then propagates down the axon, invades the synaptic terminal, and leads to the release of chemical transmitter. The transmitter diffuses across the cleft and binds with receptor sites on the postsynaptic membrane to trigger the illustrated sequence of potential changes. Note that there is a distinct delay between the application of the electrical stimulus and the production of any potential change in the muscle cell. The delay is due to two factors. First, it takes time for the action potential to propagate from its site of initiation down the motor axon. Second, there is a delay due to the time necessary for the chemical transmitter substance to be released from the presynaptic terminal, diffuse across the synaptic cleft, and produce the permeability changes that trigger the potential changes recorded in the postsynaptic muscle cell.

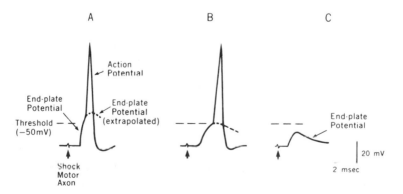

FIG. 12.5. The recording of the end-plate potential (EPP) and the effects of curare. **(A)** Normal. **(B)** Low dose of curare. **(C)** High dose of curare.

There are two components to the potential changes in the muscle that are produced as a result of stimulating the motor axon. At first, there is a relatively slow rising potential. At a potential of about -50 mV, there is a sharp inflection where a second potential is triggered. This second potential, the action potential, quickly reaches a peak value and then rapidly decays back to the resting potential. In the present discussion, the major focus is not the action potential but the somewhat slower initial underlying event that triggers the action potential.

An interesting chemical substance that has facilitated the analysis of synaptic transmission at the neuromuscular junction is curare. (Curare is derived from plants and is used by some South American Indians for arrow poison.) If a low dose of curare is added (Fig. 12.5B) and the motor axon is again stimulated, the slower underlying event is reduced in amplitude but is still capable of depolarizing the muscle cell to threshold and initiating an action potential (assume that threshold in this cell is about -50 mV). If a somewhat higher dose of curare is added, the slower underlying event is reduced further (Fig. 12.5C) and now fails to reach threshold. This underlying postsynaptic potential that is produced as a result of stimulation of the presynaptic motor axon and release of chemical transmitter substance is known as the end-plate potential (EPP). It is called the end-plate potential because it is the potential that is recorded at the motor end plate. One of the clear results of this experiment is that curare reduces the amplitude of the EPP. If sufficient curare is added, the EPP is completely abolished. A person poisoned with curare will asphyxiate because curare blocks neuromuscular transmission at the respiratory muscles. Note that if the muscle cell is artificially depolarized to threshold in the presence of a high dose of curare, an action potential can still be initiated that is indistinguishable from an action potential produced artificially in the absence of curare. Thus, although curare is effective in blocking the EPP, it has no direct effect on the action potential. Since curare does not affect the action potential, it does not affect the voltage-dependent Na^+ and K^+ channels that underlie the action potential.

The EPP is the critical event underlying the initiation of an action potential in a muscle cell. For this reason, we will explore some of the properties of EPPs in considerable detail.

PROPAGATION OF THE END-PLATE POTENTIAL

One obvious question is whether the EPP is propagated in an all-or-nothing fashion like the action potential. Figure 12.6 shows a simple experiment designed to answer this question. In this experiment the motor axon is electrically stimulated, and multiple intracellular recordings are made from the muscle cell at 1-mm intervals from the end plate. There is a large EPP (V_o) at the region of the motor end plate. However, at more distant regions from the end plate, the amplitude of the EPP becomes smaller. In fact, the decay is an exponential function of distance. This indicates that the EPP is not propagated in an all-or-nothing fashion. Rather, it spreads with decrement, just as weak artificially produced hyperpolarizations or

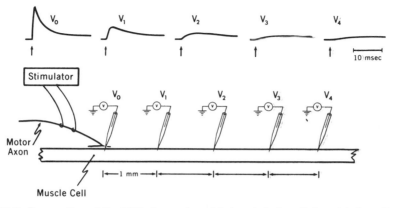

FIG. 12.6. Propagation of the EPP. A muscle cell is impaled at multiple points from the end plate, and the motor axon is stimulated. The amplitude of the EPP is maximum at the motor end plate. As recordings are made more distant from the end plate, the amplitude of the EPP becomes smaller. (Modified from P. Fatt and B. Katz, *J Physiol* 1951;115:320–370.)

depolarizations (subthreshold) spread with decrement along a nerve axon. In Fig. 12.6, an agent such as curare is utilized so that the EPP is reduced in size and fails to trigger an action potential. When the EPP triggers an action potential (as occurs in the absence of curare), the action potential propagates without decrement along the muscle cell.

ACETYLCHOLINE HYPOTHESIS FOR SYNAPTIC TRANSMISSION AT THE NEUROMUSCULAR JUNCTION

One of the major breakthroughs in the characterization of the sequence of events that underlie synaptic transmission at the neuromuscular junction was the identification of the chemical transmitter substance. In 1936, Dale and his colleagues found that electrical stimulation of motor axons led to an increase in the concentration of a substance in the perfusing solution, which they identified as acetylcholine (ACh). When the isolated ACh was injected into the arterial supply it was capable of producing large muscular contractions. Subsequent studies on the transmitter substance utilized by the neuromuscular junction have confirmed and greatly extended the original observations. As a result, it is now possible to describe the total sequence of events underlying synaptic transmission at the neuromuscular junction (Fig. 12.7).

Acetylcholine is synthesized and stored in the presynaptic terminals of motor axons. As a result of a nerve action potential that invades the presynaptic terminal, ACh is released into the synaptic cleft. Acetylcholine diffuses across the synaptic cleft and combines with receptors on the postsynaptic or the postjunctional membrane. When ACh binds with these receptors, there is an increase in Na^+ and K^+ permeabilities. The increase in Na^+ and K^+ permeabilities depolarizes the postjunctional, or the postsynaptic, membrane. This depolarization is known as the

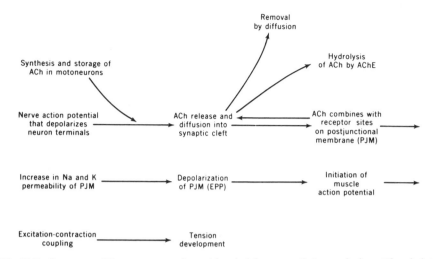

FIG. 12.7. Summary of the sequence of events underlying synaptic transmission at the skeletal neuromuscular junction: (PJM) postjunctional membrane; (EPP) end-plate potential; (ACh) acetylcholine; (AChE) acetylcholinesterase.

EPP. Normally, the EPP is approximately 50 mV in amplitude, so the threshold level of the muscle is easily reached, and an action potential is triggered. The action potential in the muscle cell leads to muscular contraction. The EPP is a transient event that persists for about 10 msec. There are two reasons for its transient nature. First, ACh diffuses away from the synaptic cleft and produces no further permeability changes. Second, there is a substance in the synaptic cleft known as acetylcholinesterase (AChE). AChE hydrolyzes, or breaks down, ACh into inactive substances. In the following sections this sequence of events is examined in greater detail.

Role of AChE

One aspect of the ACh hypothesis for chemical synaptic transmission at the neuromuscular junction involves the action of AChE. If AChE acts to break down ACh into inactive forms, one would predict that preventing this hydrolysis would allow ACh to act in the synaptic cleft for a longer time and produce a larger and longer lasting EPP. A group of substances have been identified, one of which is known as neostigmine, that block the action of AChE. Figure 12.8 illustrates the effects of neostigmine on the EPP. The lower trace shows a normal EPP. Once again the EPP is reduced (for example, by a low dose of curare) such that the EPP fails to trigger an action potential. The upper trace shows the effect of adding neostigmine to the extracellular medium. In the presence of neostigmine, the amplitude and duration of the EPP are increased.

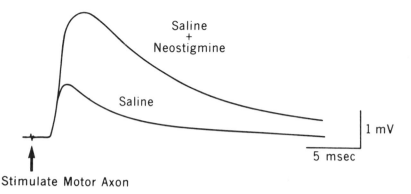

FIG. 12.8. Effects of neostigmine on the EPP. In a saline solution containing an agent such as a small dose of curare to keep the EPP subthreshold, a small EPP is produced. When neostigmine (an agent that blocks the actions of AChE) is also added to the bath, the amplitude and duration of the EPP increase.

Iontophoresis of ACh

Another prediction of the ACh hypothesis is that it should be possible to mimic the release of ACh from the presynaptic terminal by artificially applying some ACh to the vicinity of the neuromuscular junction. A simple technique called iontophoresis is available to precisely deliver very small amounts of ACh to restricted regions of a cell. The technique is illustrated in Fig. 12.9. One intracellular microelectrode is used to record the membrane potential. Another extracellular microelectrode is filled with ACh and is placed close to the neuromuscular junction (but it does not impale either the motor axon or the muscle). Acetylcholine is a positively charged molecule. If the positive pole of a battery is connected to the ACh elec-

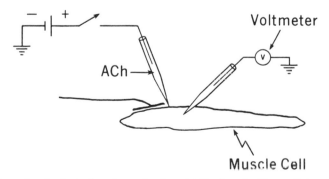

FIG. 12.9. Procedure for iontophoretic application of ACh. Standard intracellular recordings are made from a skeletal muscle cell while a second micropipette filled with ACh is positioned near the neuromuscular junction. By applying a positive voltage to the ACh-filled micropipette, small quantities of ACh can be ejected in the vicinity of the motor end plate.

trode, the positive charge of the battery will force ACh out of the electrode. The small amount of ACh ejected from the tip of the micropipette could then bind with any ACh receptors on the muscle cell. If ACh is the transmitter at the neuromuscular junction, iontophoretic ejection of ACh should produce a potential change in the muscle cell similar to that produced by stimulation of the motor axon.

Figure 12.10 illustrates the results. The upper trace is a normal EPP produced by stimulating the motor axon. The small initial downward deflection is the stimulus artifact that indicates the point in time when the motor axon is electrically stimulated. The lower trace shows the result of the iontophoretic application of ACh. The potential changes produced are nearly identical. Experiments similar to Fig. 12.10 have demonstrated a number of other aspects of synaptic transmission at the neuromuscular junction. When ACh is applied in the presence of neostigmine, the size of the potential is enhanced. These results are consistent with the cholinergic nature of synaptic transmission. If the preparation is perfused with neostigmine, ACh is not broken down by AChE, more receptors are bound, the permeability changes are greater, and the resultant potential change is greater. The iontophoretic potential is also affected by curare. By adding curare to the experimental preparation, the size of the iontophoretic response is reduced. In addition, the iontophoretically produced response is not affected by tetrodotoxin (TTX). We have already learned that TTX blocks the voltage-dependent change in Na^+ permeability underlying the action potential. Since TTX has no effect on the iontophoretic application of ACh, channels in the membrane that are sensitive to ACh must be different from the channels in the membrane that underlie the action potential. It has also been observed that injecting ACh into the muscle cell produces no potential change. Thus, the receptors for ACh must be on the outer surface of the muscle cell.

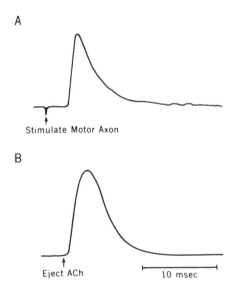

FIG. 12.10. Comparison of the EPP **(A)** and the response to iontophoretic application of ACh **(B)**.

Iontophoretic applications of ACh to points along the muscle distant from the end plate yield no potential changes. A potential change due to local application of ACh is only obtained in the immediate vicinity of the motor end plate. Presumably, there is no potential change produced by ACh at sites more distant from the motor end plate, because there are no receptors for ACh at these more distant sites. The receptors for ACh are located at the neuromuscular junction. One additional insight has been obtained from the iontophoretic application of ACh. As indicated above, curare blocks the EPP. The action of curare could be at two basic sites, however. Curare could be acting to reduce the release of chemical transmitter from the presynaptic terminal. Alternatively, it could have a postsynaptic effect. When ACh is artificially applied, the possibility of presynaptic changes is eliminated. Therefore, the reduction of the iontophoretic potential with curare is due to a postsynaptic action. As a result of these and other studies it is now known that curare is a competitive inhibitor of ACh. Curare binds with the same receptor site on the postjunctional membrane as does ACh. Although it binds with the receptors, it does not produce the resultant permeability changes.

IONIC MECHANISMS UNDERLYING THE END-PLATE POTENTIAL

How does ACh produce the permeability change responsible for the EPP? In the early 1950s, Bernard Katz and his colleagues proposed that the binding of ACh with receptors on the postjunctional membrane led to a simultaneous increase in Na^+ and K^+ permeabilities that depolarized the muscle cell toward a value of about 0 mV.

If a membrane is only permeable to Na^+ and K^+, then the Goldman-Hodgkin-Katz (GHK) equation is applicable:

$$V_m = 60 \log \frac{[K_o^+] + \alpha[Na_o^+]}{[K_i^+] + \alpha[Na_i^+]} \text{ (mV)} \qquad [12.1]$$

where $\alpha = P_{Na}/P_K$.

If $\alpha = 1$ and the K^+ and Na^+ concentrations on the inside and outside of the cell are substituted into the equation, a membrane potential of about 0 mV is predicted. Figure 12.11 illustrates one type of experiment used to test this hypothesis and to examine the ionic mechanisms underlying the EPP. The experiment once again utilizes the reduced nerve–muscle preparation. The motor axon is stimulated, and a microelectrode in the muscle cell is used to record the membrane potential and the EPP. In this case, however, there is another electrode placed in the muscle cell that is used to artificially change the membrane potential. The consequences of delivering electric shocks to the motor axon while the membrane potential of the muscle cell is systematically varied are illustrated in Fig. 12.11B. In the lower trace, the muscle cell is at its resting level of about −80 mV. The stimulation produces an EPP similar to those described above (e.g., Fig. 12.5C). The interesting observation occurs when the muscle cell is artificially depolarized to a value of about 0

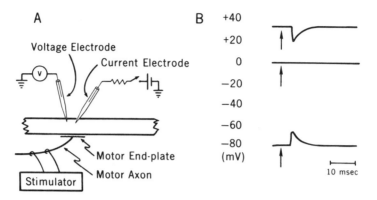

FIG. 12.11. Experimental preparation used to study the ionic mechanisms underlying the EPP.
(A) One microelectrode is used to record the membrane potential while a second is used to
artificially depolarize the cell to various levels. **(B)** At each membrane potential, the motor axon is
stimulated, and an EPP is recorded. The potential where the EPP reverses provides insight into
the possible ions that produce the response.

mV.[1] Now the identical stimulus to the motor axon produces no potential change in
the muscle. If the cell is depolarized to a value of about +30 mV and the motor
axon again stimulated, an EPP is produced but is actually reversed in sign; there is a
downward deflection. In summary, there is an upward deflection when the membrane potential is at −80 mV, no potential when the membrane is at 0 mV, and a
downward deflection when the membrane potential is moved to +30 mV. There is
a simple explanation for these results. No matter what the membrane potential, the
effect of the ACh binding with receptors is to produce a permeability change that
tends to move the membrane potential toward 0 mV. If the cell is more negative
than 0 mV, an upward deflection is recorded. If the cell is more positive than 0 mV,
a downward deflection is recorded. If the cell is at 0 mV, there is no deflection
because the cell is already at 0 mV. This 0-mV level is known as the synaptic
reversal potential, because it is the potential where the sign of the synaptic response
reverses.

Acetylcholine appears to produce a simultaneous and approximately equal increase in Na^+ and K^+ permeabilities. As a result of the equal permeability to Na^+
and K^+, the membrane potential tends to depolarize toward 0 mV. As the muscle
membrane depolarizes toward 0 mV, threshold is reached, and an action potential is
initiated or triggered. The muscle action potential then leads to muscular contraction. Although the effect of ACh is to depolarize the muscle cell toward 0 mV, this
value is never achieved even with optimal recording conditions (no curare, neostigmine in the bath). The simple reason for this is that the membrane channels opened
by ACh are only a small fraction of the ion channels in the muscle cell. These other
channels (not affected by ACh) tend to hold the membrane potential at the resting

[1] If the depolarization is applied slowly, no action potentials will be produced due to the process of
accommodation (see Chapter 9).

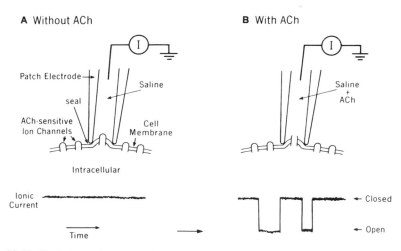

FIG. 12.12. Techniques for measuring the conductance change produced by the opening of single ACh-sensitive membrane channels. **(A)** When the solution in the patch electrode contains no ACh, the membrane channel is closed. **(B)** With ACh in the pipette the channel opens and closes in a statistical fashion. (Modified from E. Neher and B. Sakmann, *Nature* 1976;260:779–802.)

potential and prevent the membrane potential from reaching the 0-mV level. When the EPP triggers an action potential, a potential more positive than 0 mV is reached, but this is a reflection of the ionic mechanisms underlying the action potential and not the ionic mechanisms underlying the EPP.

During the past fifteen years, considerable evidence has accumulated about the molecular events associated with the actions of ACh. Using patch recording techniques (Fig. 12.12), it has been possible to measure the ionic current flowing through single channels opened by ACh. When the micropipette contains normal saline without ACh, no electrical changes are observed. However, when ACh is added to the electrode, small steplike fluctuations are observed, which are the result of ions flowing through specific membrane channels that are opened by ACh. The electrical events associated with the opening of a single channel are extremely small, and, as a result, any single channel makes a small contribution to the normal EPP. As a result of these patch recording techniques, three general conclusions can be drawn. First, ACh causes the opening of individual ionic channels (for a channel to open, two molecules of ACh must bind to the receptor). Second, when an ACh-sensitive channel opens, it does so in an all-or-nothing fashion. Increasing the concentration of ACh in the patch microelectrode does not increase the permeability of the channel; rather, it increases its probability of opening. Third, when a larger region of the muscle, and thus more than one channel, is exposed to ACh the net permeability is larger because more individual channels are opened, each in their characteristic all-or-nothing fashion. It is this summation of the permeability of many individual open channels that gives rise to the normal EPP. The properties of ACh-sensitive channels and voltage-sensitive Na^+ channels are similar in that both channels open in an all-or-nothing fashion, and as a result, the macroscopic effects that are recorded are due to the summation of many individual open ion channels. The two types of channels differ, however, in that one is opened by a chemical agent, whereas the other is opened by depolarization.

Structure of the Ligand-Gated ACh Receptor

Unlike the Na^+ channel, which consists of a single polypeptide chain with four homologous domains (each of which has six membrane-spanning regions, see. Fig. 9.9), the cation channel of the ACh receptor consists of five subunits, four of which are from separate gene products. The subunits have been designated α, β, γ, and δ, and are arranged in the stoichiometry $\alpha_2\beta\gamma\delta$ (Fig. 12.13A,B). The subunits are approximately 50% homologous at the amino acid level indicating that they evolved from a common ancestral gene. A common feature to all subunits is their four membrane-spanning regions (designated TM1 to TM4; Fig. 12.13B,C). The pore of the channel is believed to be formed by the proper alignment of the TM2 regions of each of the five subunits (Fig. 12.13B). A key difference between the α subunit and the other subunits is its extended amino terminal (NH_2) domain that contains the extracellular binding site for ACh (Fig. 12.13C). The long cytoplasmic loop found between the TM3 and TM4 regions is present in all of the ACh subunits. It may function as a binding site to the cytoskeleton.

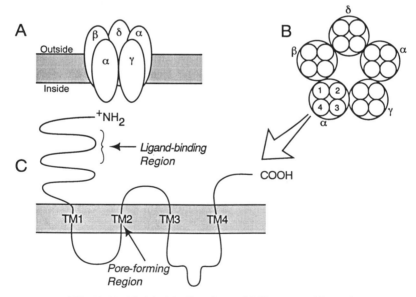

FIG. 12.13. Model of the ligand-gated ACh receptor/channel.

PRESYNAPTIC EVENTS UNDERLYING THE RELEASE OF NEUROTRANSMITTER

Voltage-Dependent Release of Neurotransmitter

One of the most interesting aspects of synaptic transmission is the mechanism by which an action potential in the presynaptic terminal triggers the release of the chemical transmitter substance. One possibility is that transmitter release is due to some aspect of the sequence of permeability changes underlying the action potential in the presynaptic terminal. The action potential is associated with a slight influx of Na^+ and a slight efflux of K^+. Perhaps in some way either the small influx of Na^+ or efflux of K^+ disturbs the intracellular environment and causes the release of transmitter.

Figure 12.14 illustrates an experiment to examine this hypothesis. Rather than using the neuromuscular junction, a more advantageous experimental preparation, the squid giant synapse is used. The squid giant synapse is so large that the presynaptic terminal can be impaled with two electrodes: one to record the presynaptic potential and the other to depolarize the terminal artificially (Fig. 12.14) (this is not possible at the skeletal neuromuscular junction). A third electrode is used to record the potential changes in the postsynaptic cell. To examine the possible roles of Na^+ influx and K^+ efflux in triggering release, the preparation is exposed to TTX (to block Na^+ influx) and tetraethylammonium (TEA) (to block K^+ efflux). By passing brief current pulses of various amplitudes into the stimulating electrode, the membrane potential of the presynaptic terminal can be depolarized to a variety of different membrane potentials without initiating action potentials (Fig. 12.14). When the presynaptic cell is depolarized by a small amount, there is no postsynaptic potential. A striking observation is made, however, when greater levels of depolarization are applied. In this case, postsynaptic potentials are produced, and their amplitudes are dependent on the level of depolarization. Thus, this experiment indicates that transmitter release is voltage-dependent and that artificial depolarization of the presynaptic terminal is able to produce a postsynaptic potential. Because the preparation has been treated with TTX and TEA, this experiment clearly demonstrates that the voltage-dependent changes in Na^+ and K^+ permeability that underlie the action potential have no direct effect on the release of chemical transmitter substance from the presynaptic terminal. One merely has to depolarize the cell to get the release of transmitter. Normally, the action potential is essential because it is the vehicle by which the terminal is depolarized. Although changes in Na^+ and K^+ permeability are important for producing an action potential, the direct influx of Na^+ and efflux of K^+ are, by themselves, not causally related to the release of the chemical transmitter substance.

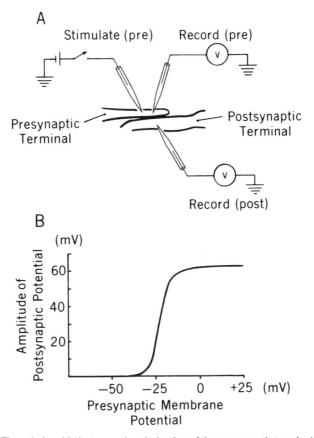

FIG. 12.14. The relationship between depolarization of the presynaptic terminal and the release of neurotransmitter. **(A)** Experimental setup. **(B)** As the presynaptic terminal is depolarized to greater levels, transmitter release (as measured by the amplitude of the postsynaptic potential) increases. Since the results were obtained in the presence of TEA and TTX, these results indicate that the Na^+ influx or K^+ efflux that occurs during the action potential is not necessary for the release of neurotransmitter. (Modified from B. Katz and R. Miledi, *J Physiol* 1967;192:407–436.)

Role of Ca^{2+} in Transmitter Release

If Na^+ and K^+ are not directly involved in the release of chemical transmitter, might other ions be involved? It has been known for many years that Ca^{2+} in the extracellular medium is important for chemical transmission. If the concentration of Ca^{2+} in the extracellular medium is decreased, synaptic transmission is reduced (increasing the concentration of Mg^{2+} in the extracellular medium also reduces transmitter release).

An experiment that examines the role of Ca^{2+} in the release of chemical transmitter substances is illustrated in Figure 12.15. Here the neuromuscular junction is

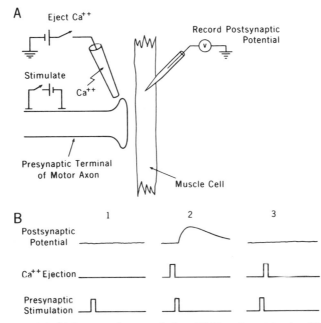

FIG. 12.15. Role of Ca^{2+} in synaptic transmission. **(A)** Experimental setup **(B)** (1) In a Ca^{2+}-free medium a depolarization of the presynaptic terminal does not lead to any transmitter release; (2) when a small amount of Ca^{2+} is ejected just prior to a second depolarization, a postsynaptic potential is produced; (3) if Ca^{2+} is ejected after the depolarization, no potential is produced. (Modified from B. Katz and R. Miledi, *J Physiol* 1967;189:535–544.)

used, and one electrode is placed in the muscle cell to record the postsynaptic potential, and stimulating electrodes are placed close to the presynaptic terminal but not inside the terminal. The stimulating electrodes are sufficiently close to the presynaptic terminal to depolarize the terminal by extracellular means. A third electrode filled with $CaCl_2$ is positioned near the presynaptic terminal. The preparation is perfused with a Ca^{2+}-free medium. Figure 12.15B illustrates the results. In panel 1, stimulation of the presynaptic terminal produces no postsynaptic potential. Thus, in a Ca^{2+}-free medium, transmitter release is abolished. In panel 2 (Fig. 12.15B), the presynaptic stimulation is preceded by a brief ejection of Ca^{2+} from the electrode containing $CaCl_2$. Calcium ions are positively charged, so it is possible to eject a small amount of Ca^{2+} from the Ca^{2+} electrode by simply connecting the electrode to the positive pole of a battery. As a result of the brief prior ejection of Ca^{2+}, the depolarizing stimulus now produces a postsynaptic potential in the muscle cell. If Ca^{2+} is ejected after the depolarization of the presynaptic terminal, no postsynaptic potential is produced in the muscle cell (panel 3, Fig. 12.15B). This experiment clearly indicates that Ca^{2+} is absolutely essential for the release of chemical transmitters. Furthermore, it illustrates that Ca^{2+} must be present just prior to (or during) the depolarization of the presynaptic terminal. Based on these

SYNAPTIC TRANSMISSION

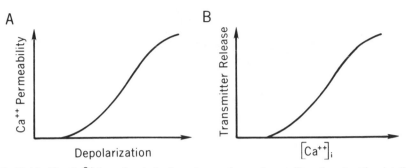

FIG. 12.16. The Ca^{2+} hypothesis for the release of neurotransmitters (see text for details).

and other experiments, Katz and his colleagues proposed the Ca^{2+} hypothesis for chemical transmitter release.

Calcium ions are in high concentration outside the cell and in low concentration inside the cell. Furthermore, the inside of the cell is negatively charged with respect to the outside. As a result, there is a large chemical and electrical driving force for the influx of Ca^{2+}. Normally, the cell is relatively impermeable to Ca^{2+}, so despite the fact that there is a large driving force, Ca^{2+} does not enter the cell. It is proposed that there is a voltage-dependent change in Ca^{2+} permeability (Fig. 12.16A). Thus, the depolarization of the presynaptic membrane results in an increase in Ca^{2+} permeability, and Ca^{2+} moves down its electrical and chemical gradient and flows into the synaptic terminal. The resultant elevation of intracellular Ca^{2+} concentration leads to the release of chemical transmitter (Fig. 12.16B).

Thus, according to this hypothesis, Ca^{2+} is the critical trigger for the release of chemical transmitter. One can artificially depolarize the presynaptic terminal to produce a voltage-dependent change in Ca^{2+} influx and release of transmitter. Normally, however, the action potential invades the terminal, depolarizes it, and leads to the subsequent voltage-dependent increase in Ca^{2+} permeability, influx of Ca^{2+}, and chemical transmitter release.

Quantal Nature of ACh Release

What is the mechanism by which Ca^{2+} causes the release of chemical transmitter? Figure 12.17 illustrates an experiment that provided some initial insights. As discussed above, the EPP is a relatively large event, 50 mV or so in amplitude under normal situations. Most scientists who had been recording the EPP at the neuromuscular junction were doing so with the amplification of their oscilloscope set at a relatively low level. Katz and his colleagues increased the amplification and examined the background noise in the absence of any stimulation. (This would be equivalent to tuning your stereo receiver to a portion of the band that has no station and maximally increasing the volume to listen to the noise.) Figure 12.17 illustrates the type of electrical measurements made at the neuromuscular junction under these conditions. The traces in Fig. 12.17A are a continuous recording from the same experiment going from left to right and down the illustration. (It is important to keep in mind that in this experiment the motor axon is not stimulated and that the recordings are made from the unstimulated muscle cell.) It is evident that in the absence of stimulation there is no absence of activity. Indeed, numerous small voltage deflections [on the average only about ½ mV in amplitude (Fig. 12.17B)] are observed. They are small compared to the relatively large potential changes observed in response to a presynaptic action potential. The potentials occur in a random fashion at an average rate of about once every 50 msec.

The small potential changes have some interesting features. When the recording electrode is placed in regions distant from the motor end plate, the potentials disappear. For this reason, these potential changes are called miniature EPPs or MEPPs. MEPPs, because they are EPPs that occurred in the vicinity of the motor end plate and are much smaller than the EPP normally produced by stimulating the motor

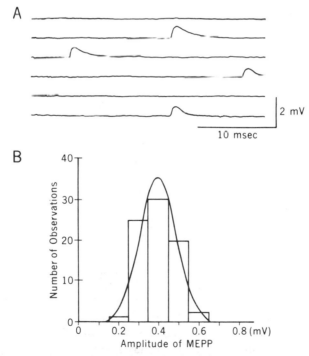

FIG. 12.17. The miniature end-plate potential (MEPP). **(A)** In the absence of nerve stimulation, small spontaneously occurring potentials are recorded. The *successive traces* are a continuous recording from a neuromuscular junction. The MEPPs are approximately 0.4 mV in amplitude and occur about once every 50 msec. (Modified from A. W. Liley, *J Physiol* 1956;133:571–587.) **(B)** Histogram representing the number of times a MEPP of a given amplitude is recorded. There is a variation in amplitude, but the MEPPs appear to be derived from a single population and have an average amplitude of 0.4 mV. (Modified from I. A. Boyd and A. R. Martin, *J Physiol* 1956;132:74–91.)

axon. If curare is added to the extracellular medium, the size of the MEPPs becomes smaller. If neostigmine is added to the extracellular medium, the size of the MEPPs becomes larger. Based on these considerations, it appears that MEPPs are due to the spontaneous and random release of ACh from the presynaptic terminal. Acetylcholine presumably is in relatively high concentration in the presynaptic terminal. So, by chance some of that ACh might diffuse out of the presynaptic terminal, bind with postsynaptic receptors, and produce small permeability changes.

How much ACh is necessary to actually produce a MEPP? The simplest answer is that a MEPP is produced by one ACh molecule binding to a receptor and opening a single ACh-sensitive channel. When small amounts of ACh are ejected into the vicinity of the neuromuscular junction, however, potentials much smaller than MEPPs are produced. Indeed, the single-channel recording techniques (e.g., Fig. 12.12) indicate that the opening of a single ACh-sensitive channel produces a potential change of approximately 0.4 μV. This indicates that the MEPPs are not due to the release of a single molecule of ACh. Because the opening of a single ACh-sensitive channel produces a potential change of approximately 0.4 μV, it would take at least 1,000 molecules of ACh to produce a potential change the size of one MEPP (0.4 mV). Indeed, because of loss due to diffusion in the synaptic cleft and the fact that two ACh molecules are necessary to open a channel, approximately 10^4 ACh molecules are required to produce a MEPP. Therefore, it appears that it is not a single molecule of ACh that is spontaneously and randomly released from the presynaptic terminal, but a package of 10^4 molecules of ACh. It is now fairly well established that the morphological locus of this package of 10^4 ACh molecules is the synaptic vesicle found in high concentration in a presynaptic terminal. Chemical analyses have revealed that the synaptic vesicles in the motor axon contain ACh.

Is the normal EPP due to the release of these packages of ACh as well? One observation consistent with this hypothesis is that when Ca^{2+} is injected into the presynaptic terminal, the frequency of occurrence of the MEPPs is increased. This indicates that the release of the vesicles is Ca^{2+}-dependent. As a result, it is attractive to think that when a larger amount of Ca^{2+} enters the presynaptic terminal during an action potential, a large number of vesicles is released synchronously. Figure 12.18 illustrates a test of this hypothesis. To make the interpretation of the experiment easier, the Ca^{2+} concentration in the extracellular medium is reduced to a point where the evoked EPPs are quite small (approximately the same size as the MEPPs). The left side of the traces in Fig. 12.18A illustrate some of the spontaneous MEPPs [approximately 0.4 mV in amplitude (asterisks)]. The small vertical lines on the traces illustrate the point in time at which the stimulus is applied to the motor axon. When the motor axon is stimulated in the low-Ca^{2+} solution, very small EPPs are produced, but there is considerable variation in their size. For example, sometimes the EPP is about the same size as the MEPP (spontaneously occurring) (trace 3, Fig. 12.18A). Such a response is called a unit EPP. Sometimes no EPP is produced by the stimulus (traces 2 and 6, Fig. 12.18A) and at other times, an EPP is produced that is about twice the size of the unit (double) (traces 4, 7, and 8, Fig. 12.18A). In other cases, EPPs are produced that are about three times the size

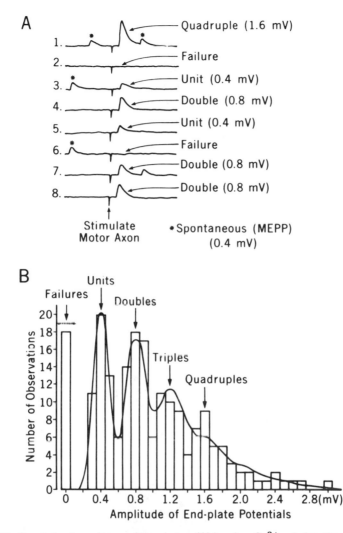

FIG. 12.18. Quantal nature of transmitter release. **(A)** In a low-Ca^{2+} solution, the motor axon is repeatedly stimulated, and the amplitude of the EPP is monitored. In such a solution the amplitude of the EPP is small and variable. *Asterisks* indicate spontaneous MEPPs. (Modified from A. W. Lⅼley, *J Physiol* 1956;133:571–587.) **(B)** A histogram showing the number of times EPP of various amplitudes are recorded in the low-Ca^{2+} solution. There are a large number of EPPs that have an amplitude identical to the amplitude of the MEPP (0.4 mV). Other peaks are multiples of the first. (Modified from I. A. Boyd and A. R. Martin, *J Physiol* 1956;132:74–91.)

of the unit or the spontaneous EPP. Sometimes there is an EPP produced that is about four times the size of the unit EPP, which is called a quadruple (trace 1, Fig. 12.18A). By measuring the amplitude of all of the evoked EPPs, a plot can be made of the number of times the EPPs of a given amplitude are observed. Similarly, an amplitude distribution of the spontaneous MEPPs can be made (e.g., Fig. 12.17B).

As we have already learned, the spontaneous MEPPs have an average value of approximately 0.4 mV. The plot of the evoked EPPs produced in low Ca^{2+} reveals a remarkable finding. Namely, there is a multimodal distribution of sizes (Fig. 12.18B). There is a large proportion of EPPs that have an average value of approximately 0.4 mV, another large percentage that have an average value of approximately 0.8 mV, and some that have an average value of approximately 1.2 mV. The critical finding is that the size of the first peak of the evoked EPP (0.4 mV) is nearly identical to the size of the spontaneous MEPPs and that subsequent peaks are multiples of the first.

Based on these observations, Katz and his colleagues proposed the quantal nature of transmitter release. The hypothesis was developed at the neuromuscular junction, but the quantal nature of release appears to be widely applicable to chemical synapses. Transmitter release at some synapses may be nonvesicular, however. According to the quantal hypothesis, the size of the EPP fluctuates because different numbers of packages (quanta) of ACh are released with each stimulus. The smallest EPP is due to the release of a single package of ACh containing 10^4 molecules. An EPP twice that size is due to the release of two packages, and so forth. The EPP of about 50 mV produced in normal concentrations of Ca^{2+} is a compound of the individual contributions of many packages of ACh. Since each package of ACh produces an amplitude of approximately 0.4 mV, the normal EPP is due to the simultaneous release of more than 100 packages of ACh, each containing 10^4 molecules and each quanta producing a potential change by itself of 0.4 mV. Calcium ions appear to be the critical step in causing the release of the vesicles. In low extracellular Ca^{2+}, an action potential in the presynaptic terminal results in a small amount of Ca^{2+} influx and a small number of vesicles released. When the extra-

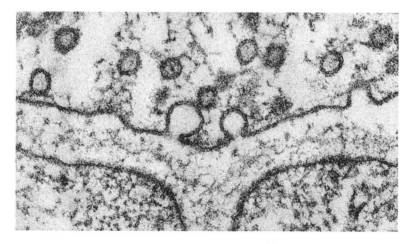

FIG. 12.19. Exocytosis. Transmitter vesicles "caught" in the process of exocytosis. The top portion of the photograph is the presynaptic ending. (Micrograph produced by Dr. John E. Heuser of Washington University School of Medicine, St. Louis, MO.)

cellular Ca^{2+} concentration is high, a large amount of Ca^{2+} flows into the terminal and causes a larger number of vesicles to be released (and therefore a larger EPP). The Ca^{2+} that enters the cell promotes the binding of vesicles with the inside membrane of the presynaptic terminal to promote exocytosis (Fig. 12.19). Considerable progress has been made in identifying the specific proteins involved in the docking, fusion, and regulation of vesicular release.

SUMMARY OF THE SEQUENCE OF EVENTS UNDERLYING SYNAPTIC TRANSMISSION AT THE SKELETAL NEUROMUSCULAR JUNCTION

As a result of work during the past several decades, a fairly complete understanding of the sequence of events underlying synaptic transmission at the skeletal neuromuscular junction is available (Fig. 12.20; see also Table 12.1). The resting potential at both the presynaptic terminal of the motor axon and the postsynaptic muscle cell is due to the high resting K^+ permeability. At rest, a small amount of transmitter release occurs due to the spontaneous release of vesicles (quanta) from the presynaptic terminal. The spontaneous release is presumably due to the low basal levels of Ca^{2+} in the presynaptic terminal. The resultant MEPPs are only approximately 0.4 mV and are insufficient to trigger an action potential in the muscle cell. When an action potential propagates down the ventral root, it eventually invades the

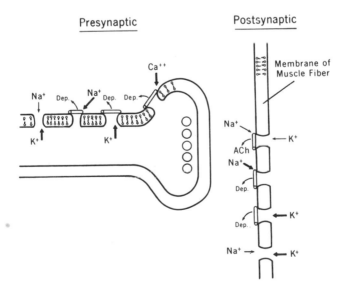

FIG. 12.20. Summary diagram of the membrane channels important for the process of synaptic transmission at the skeletal neuromuscular junction. Dep., depolarization.

TABLE 12.1. *Properties of action potentials and synaptic potentials at the skeletal neuromuscular junction*

	Synaptic potential (end-plate potential)	Action potential
Changes in membrane conductance	Initiated by ACh	Initiated by depolarization (the end-plate potential)
During rising phase (initiation)	Simultaneous chemically gated increase in g_{Na} and g_K	Specific voltage-dependent increase in g_{Na}
During falling phase	Passive decay; diffusion and AChE	Specific increase in g_K and decrease in g_{Na}
Duration	10–20 msec	1–3 msec
Equilibrium potential	Reversal potential close to 0 mV	E_{Na} (+55 mV)
Pharmacology	Blocked by curare; enhanced by neostigmine; not blocked by TTX	Blocked by TTX, but not by curare; not affected by neostigmine
Propagation	Propagates with decrement	All-or-nothing
Other features	No evidence for regenerative action or refractory period	Regenerative rise, followed by absolute and relative refractory periods

presynaptic terminal. As the action potential invades the terminal, the terminal is depolarized, and that depolarization leads to the opening of voltage-dependent Ca^{2+} channels. Calcium ions move down their electrochemical gradient and enter the terminal. The resultant increase in the intracellular Ca^{2+} concentration leads to the release of 100 or so synaptic vesicles, each containing 10^4 molecules of ACh. The released ACh diffuses across the synaptic cleft and binds to receptors on the postsynaptic membrane. The binding of ACh with ACh-sensitive channels causes individual channels that are normally closed or open in an all-or-nothing fashion. The chemically gated channels opened by ACh are distinct from the voltage-gated channels (Na^+ and K^+) that underlie the action potential in the axon and skeletal muscle cell. The channel opened by ACh is equally permeable to Na^+ and K^+, and as a result the postsynaptic membrane is depolarized. This postsynaptic potential (the EPP), if sufficiently large (as it normally is), brings the muscle membrane potential to threshold. A new potential, the action potential, is initiated in the muscle cell by the voltage-dependent changes in Na^+ and K^+ permeabilities. The muscle action potential propagates along the muscle cell membrane and leads, by a process known as excitation-concentration coupling, to the development of muscle tension.

MYASTHENIA GRAVIS

Myasthenia gravis is a debilitating neuromuscular disease associated with weakness and fatigability of skeletal muscle. The condition is aggravated by exercise. The muscular weakness and fatigability are in turn associated with end-plate potentials that are smaller in amplitude than normal. Recall that there is generally a one-to-one relationship between an action potential in a motor axon and an action potential in the skeletal muscle cell. This is so because the EPP is normally quite large (50 mV in amplitude). As a result, an EPP in a muscle cell with a resting potential of approximately -80 mV is always capable of depolarizing the muscle cell past threshold (approximately -50 mV) and initiating a skeletal muscle action potential. Note that to reach threshold, an EPP need only have an amplitude of 30 mV (e.g., -80 mV $+30$ mV $= -50$ mV). Thus, there is normally a "safety factor" of approximately 20 mV. In myasthenia gravis, the EPP is smaller, and there is less of a safety factor. In severe cases, the EPP is so small that it fails to reach threshold, and no muscular contraction is produced.

The fatigability associated with myasthenia gravis is explained by the tendency for transmitter release to depress with repeated activation of a motor neuron. Such depression may be due to depletion of the pool of synaptic vesicles in the presynaptic terminal. In normal healthy adults, depression of transmitter release is of little consequence, since the safety factor is so large. For example, the one-to-one relationship between motor neuron activation and skeletal muscle contraction can be maintained even if the EPP is reduced from an amplitude of 50 mV to an amplitude of 30 mV. In patients with myasthenia gravis, however, the EPP is already reduced. As a result, when the motor neuron is repeatedly activated and transmitter release is depressed, the EPP becomes subthreshold and fails to elicit an action potential in the muscle cell. Thus, while initial motor responses in the myasthenic patient may be relatively normal, they fatigue rapidly as the EPP falls below threshold.

At the molecular level it is now known that the reduction of the EPP in patients with myasthenia gravis is due to reduction in the number of ACh receptors in the postjunctional membrane. Because of a reduced number of receptors, the postsynaptic permeability changes and the EPP are smaller. Current evidence indicates that the reduction of ACh receptors is due to an autoimmune response to the ACh receptor.

There is no known cure for myasthenia gravis. A common treatment, however, is the use of neostigmine. Neostigmine, by blocking the actions of AChE, makes more ACh available to bind with postjunctional ACh receptors and thus partially compensates for the reduced number of receptors in the myasthenic patient.

TETANY

Tetany is a pathological condition accompanying hypocalcemia (low extracellular Ca^{2+}) that is associated with hyperexcitability of nerve and muscle cells. We have just learned that low Ca^{2+} tends to reduce chemical transmitter release, but low Ca^{2+} also has an effect on the postsynaptic cells as well (in this case the muscle cell). Lowering the Ca^{2+} concentration tends to reduce the threshold for initiating action potentials. The combined effect of low Ca^{2+} is to make the membrane of the muscle cell more easy to depolarize and thus better able to initiate action potentials, despite the decreased chemical transmitter release.

BIBLIOGRAPHY

Aidley DJ. *The physiology of excitable cells*, 2nd ed. Cambridge: Cambridge University Press, 1978; Chapters 6–8.

Hille B. *Ionic channels of excitable membranes*, 2nd ed. Sunderland, MA: Sinauer Associates; 1992.

Kandel ER, Schwartz JH, Jessell TM. *Principles of neural science*, 3rd ed. New York: Elsevier/North-Holland, 1991; Chapters 10, and 13.

Katz B. *The release of neural transmitter substances*. Springfield: Charles C Thomas, 1969.

Nicholls JG, Martin AR, Wallace BG. *From neuron to brain*, 3rd ed. Sunderland, MA; Sinauer Associates, 1992: Chapter 7.

Schmidt RF, ed. *Fundamentals of neurophysiology*, 2nd ed. New York: Springer-Verlag, 1978; Chapters 3 and 4.

ADDITIONAL READING

Attwell D, Barbour B, Szatkowski M. Nonvesicular release of neurotransmitter. *Neuron* 1993;11:401–407.

Bennett MVL. Electrical transmission: a functional analysis and comparison to chemical transmission, In: Kandel ER, ed. *Handbook of physiology, Section 1: The nervous system*, vol. 1, pt 1. Bethesda: American Physiological Society, 1977:357–416.

Bennet MK, Scheller RH. A molecular description of synaptic vesicle membrane trafficking. *Annu Rev Biochem* 1994;63:63–100.

Boyd IA, Martin AR. The end-plate potential in mammalian muscle. *J Physiol* 1956;132:74–91.

del Castillo J, Katz B: Quantal components of the end-plate potential. *J Physiol* 1954;124:560–573.

del Castillo J, Katz B. Statistical factors involved in neuromuscular facilitation and depression. *J Physiol* 1954;124:574–585.

Dermietzel R, Spray DC. Gap junctions in the brain: where, what type, how many and why? *TINS* 1993; 16:186–192.

Fatt P, Katz B. An analysis of the end-plate potential recorded with an intracellular electrode. *J Physiol* 1951;115:320–370.

Galzi J-L, Revah F, Beisis A, Changeux J-P. Functional architecture of the nicotinic acetylcholine receptor: from electric organ to brain. *Annu Rev Pharmacol* 1991;31:37–72.

Greengard P, Valtorta F, Czernik AJ, Benfenati F. Synaptic vesicle phosphoproteins and regulation of synaptic function. *Science* 1993;259:780–786.

Katz B, Miledi R. The release of acetylcholine from nerve endings by graded electrical pulses. *Proc R Soc Lond (Biol)* 1967;167:23–38.

Katz B, Miledi R. The timing of calcium action during neuromuscular transmission. *J Physiol* 1967; 189:535–544.

Katz B, Miledi R. A study of synaptic transmission in the absence of nerve impulses. *J Physiol* 1967; 192:407–436.

Lester HA. The response to acetylcholine. *Sci Am* 1977;236:106.

Liley AW. The quantal components of the mammalian end-plate potential. *J Physiol* 1956;133:571–587.

Martin AR. Junctional transmission II. Presynaptic mechanisms, In: Kandel ER, ed. *Handbook of physiology, Section 1: The nervous system*, vol. 1, pt 1. Bethesda: American Physiological Society, 1977: 329–355.

Neher E, Sakmann B. Single-channel currents recorded from membrane of denervated frog muscle fibres. *Nature* 1976;260:779–802.

Schuetze S. Understanding channel gating: patch-clamp recordings from cloned acetylcholine receptors. *Trends Neurosci* 1986;9:140–141.

Stroud RM, Finer-Moore J. Acetylcholine receptor structure, function, and evolution. *Annu Rev Cell Biol* 1985;1:317–351.

Takeuchi A. Junctional transmission I. Postsynaptic mechanisms, In: Kandel ER, ed. *Handbook of physiology, Section 1: The nervous system*, vol. 1, pt 1. Bethesda: American Physiological Society, 1977:295–327.

Walch-Solimena C, Jahn R, Südhof TC. Synaptic vesicle proteins in exocytosis: what do we know? *Curr Opin Neurobiol* 1993;3:329–336.

Zucker RS. The role of calcium in regulating neurotransmitter release in the squid giant synapse, In: Feigenbaum, J and Hanani, M, eds. *Presynaptic regulation of neurotransmitter release: A handbook*. London: Freund Publishing House 1991;1:153–195.

13

Synaptic Transmission in the Central Nervous System

The study of synaptic transmission in the central nervous system (CNS) is both an opportunity to learn more about the diversity and richness of mechanisms underlying this process, as well as an opportunity to learn how some of the fundamental signaling properties of the nervous system, such as action potentials and synaptic potentials, work together to process information and generate behavior.

One of the simplest behaviors that is controlled by the central nervous system is the knee-jerk or stretch reflex. The tap of a neurologist's hammer to a ligament elicits a reflex extension of the leg illustrated in Fig. 13.1. The brief stretch of the

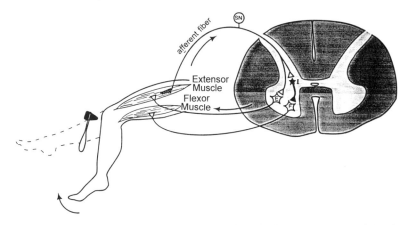

FIG. 13.1. Features of the vertebrate stretch reflex. Stretch of an extensor muscle leads to the initiation of action potentials in the afferent terminals of specialized stretch receptors. The action potentials propagate to the spinal cord via afferent fibers (sensory neurons, *SN*). The afferents make excitatory connections with extensor motor neurons (*E*). Action potentials initiated in the extensor motor neuron propagate to the periphery and lead to the activation and subsequent contraction of the extensor muscle. The afferent fibers also activate interneurons (*I*) that inhibit the flexor motor neurons (*F*).

ligament is transmitted to the extensor muscle and is detected by specific receptors in the muscle and ligament. Action potentials initiated in the stretch receptors are propagated to the spinal cord by afferent fibers. The receptors are specialized regions of sensory neurons with somata located in the dorsal root ganglia just outside the spinal column. The axons of the afferents enter the spinal cord and make at least two types of excitatory synaptic connections. First, a synaptic connection is made to the extensor motor neuron. As the result of its synaptic activation, the motor neuron fires action potentials that propagate out of the spinal cord and ultimately invade the terminal regions of the motor axon at neuromuscular junctions (see also Chapter 12). There, ACh is released, an EPP is produced, an action potential is initiated in the muscle cell, and the muscle cell is contracted producing the reflex extension of the leg. Second, a synaptic connection is made to another group of neurons called interneurons (nerve cells interposed between one type of neuron and another). The particular interneurons activated by the afferents are inhibitory interneurons, since activation of these interneurons leads to the release of a chemical transmitter substance that inhibits the flexion motor neuron. This inhibition tends to prevent an uncoordinated (improper) movement (i.e., flexion) from occurring. The reflex system illustrated in Fig. 13.1 is also known as the monosynaptic stretch reflex, because this reflex is mediated by a single ("mono") synapse in the central nervous system.

MECHANISMS FOR EXCITATION, INHIBITION, AND INTEGRATION OF SYNAPTIC POTENTIALS

The two major objectives of this chapter are the description of properties of excitatory and inhibitory synaptic transmission in the central nervous system and the illustration of how these properties are used to process information and generate simple behavior. The stretch reflex provides a suitable model system.

Figure 13.2A illustrates procedures that can be used to examine experimentally some of the components of synaptic transmission in the reflex pathway for the stretch reflex. Intracellular recordings are made from one of the sensory neurons, the extensor and flexor motor neurons, and an inhibitory interneuron. Normally, the sensory neuron is activated by stretch to the muscle, but this step can be bypassed by simply injecting a pulse of depolarizing current of sufficient magnitude into the sensory neuron to elicit an action potential. The action potential in the sensory neuron leads to a potential change in the motor neuron known as an *excitatory postsynaptic potential* (EPSP) (Fig. 13.2B). The potential is excitatory because it increases the probability of firing an action potential in the motor neuron; it is postsynaptic, because it is a potential that is recorded on the receptive (postsynaptic) side of the synapse.

Postsynaptic potentials (PSPs) in the CNS can be divided into two broad classes based on mechanisms and duration of these potentials. One class arises from the *direct* binding of a transmitter molecule(s) with a receptor–channel complex; these

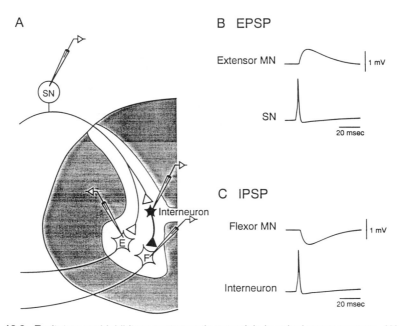

FIG. 13.2. Excitatory and inhibitory postsynaptic potentials in spinal motor neurons. **(A)** Intracellular recordings are made from a sensory neuron (*SN*), interneuron, and extensor (*E*) and flexor (*F*) motor neurons. **(B)** An action potential in the sensory neuron produces a depolarizing response in the motor neuron (*MN*). This response is called an excitatory postsynaptic potential (EPSP). **(C)** An action potential in the interneuron produces a hyperpolarizing response in the motor neuron. This response is called an inhibitory postsynaptic potential (IPSP).

receptors are called *ionotropic*. The resulting PSPs are generally short lasting and hence are sometimes called fast PSPs; they have also been referred to as *classical* due to the fact that these were the first synaptic potentials that were recorded in the CNS. Fast EPSPs also resemble the synaptic potentials at the neuromuscular junction (i.e., the EPP). One typical feature of the classical synaptic potentials is their time course. Recall that the duration of an EPP is about 20 msec. An EPSP, or an IPSP, recorded in a spinal motor neuron is of approximately the same duration.

The second class of PSPs arises from the *indirect* effect of transmitter molecule(s) binding with a receptor. For these PSPs, one common coupling mechanism is an alteration in the level of a second messenger. The receptors that produce these PSPs are called *metabotropic*. The responses can be long lasting and are therefore called slow PSPs. The mechanisms for fast PSPs mediated by ionotropic receptors will be considered first.

Postsynaptic Potentials Produced by Ionotropic Receptors

Ionic Mechanisms of EPSPs

Mechanisms responsible for fast EPSPs mediated by ionotropic receptors in the central nervous system are fairly well known. Specifically, for the synapse between the sensory neuron and the spinal motor neuron, the ionic mechanisms for the EPSP are essentially identical to the ionic mechanisms at the skeletal neuromuscular junction. Transmitter substance released from the presynaptic terminal of the sensory neuron diffuses across the synaptic cleft, binds to specific receptor sites on the postsynaptic membrane, and leads to a simultaneous increase in permeability to Na^+ and K^+, which makes the membrane potential move toward a value of 0 mV. Although this mechanism is superficially the same as that for the neuromuscular junction, two rather fundamental differences exist between the process of synaptic transmission at the sensory neuron–motor neuron synapse and the motor neuron–skeletal muscle synapse. First, these two different synapses use different transmitters. The transmitter substance at the neuromuscular junction is ACh, but the transmitter substance released by the sensory neurons is an amino acid, probably glutamate.[1] A second major difference between the sensory neuron–motor neuron synapse and the motor neuron–muscle synapse is the amplitude of the postsynaptic potential. Recall that the amplitude of the postsynaptic potential at the neuromuscular junction was approximately 50 mV and that a one-to-one relationship existed between an action potential in the spinal motor neuron and an action potential in the muscle cell. Indeed, since the EPP must only depolarize the muscle cell by approximately 30 mV to initiate an action potential, there is a safety factor of 20 mV. In contrast, the EPSP in a spinal motor neuron produced by an action potential in an afferent fiber has an amplitude of only 1 mV.

The small amplitude of the EPSP in spinal motor neurons (and other cells in the CNS) poses an interesting question. Specifically, how can an ESP with an amplitude of only 1 mV drive the membrane potential of the motor neuron (i.e., the postsynaptic neuron) to threshold and fire the spike in the motor neuron that is necessary to produce the contraction of the muscle? The answer to this question lies in the principles of temporal and spatial summation.

---

[1]Although beyond the scope of this chapter, there are many different transmitters used by the nervous system—up to 50 or more (the list grows yearly). Thus, to understand fully the process of synaptic transmission and the function of synapses in the nervous system, it is necessary to know the mechanisms for the synthesis, storage, release, and degradation or uptake, as well as the different types of receptors for each of the transmitter substances. The clinical implications of deficiencies in each of these features for each of 50 transmitters cannot be ignored. Fortunately, most of the transmitter substances can be grouped into four basic categories; these include ACh, the monoamines, the amino acids, and the peptides.

Temporal and Spatial Summation

When the ligament is stretched (Fig. 13.1), hundreds of stretch receptors are activated. Indeed, the greater the stretch, the greater the probability of activating a larger number of the stretch receptors present; this process is referred to as *recruitment*. Activation of multiple stretch receptors is not the complete story, however. Recall the principle of frequency coding in the nervous system (Chapter 8). Specifically, the greater the intensity of a stimulus, the greater the number per unit time or frequency of action potentials that is elicited in a sensory receptor. This principle applies to stretch receptors as well. Thus, the greater the stretch, the greater the number of action potentials elicited in the stretch receptor in a given interval and, therefore, the greater the number of EPSPs produced in the motor neuron from that train of action potentials in the sensory cell. Consequently, the effects of activating multiple stretch receptors add together (spatial summation), as do the effects of multiple EPSPs elicited by activation of a single stretch receptor (temporal summation). Both of these processes act in concert to depolarize the motor neuron sufficiently to elicit one or more action potentials, which then propagate to the periphery and produce the reflex.

1. *Temporal summation.* Temporal summation can be illustrated by considering the case of firing action potentials in a presynaptic neuron and monitoring the resultant EPSPs. For example in Fig. 13.3A, B, a single action potential in SN1 produces a 1-mV EPSP in the motor neuron. Two action potentials in quick succession produce two EPSPs, but note that the second EPSP occurs during the falling phase of the first, and the depolarization associated with second EPSP adds to the depolarization produced by the first. Thus, two action potentials produce a summated potential that is 2 mV in amplitude. Three action potentials in quick succession would produce a summated potential of 3 mV. In principle, thirty action potentials in quick succession would produce a potential of 30 mV and easily drive the cell to threshold. This summation is strictly a passive property of the cell. No special ionic conductance mechanisms are necessary.

A thermal analog is helpful to understand temporal summation. Consider a metal rod that has thermal properties similar to the passive electrical properties of a dendrite and the flame of a cigarette lighter that will generate heat equivalent to a depolarization generated by a synaptic current at the synapse. Now consider the consequences of presenting the flame to one end of the rod on the temperature measured in the middle of the rod. Obviously, a temperature change would be produced, and for the sake of the example, assume the temperature change is 10°C. If the flame is presented twice in quick succession, a greater temperature would be recorded, because the heat generated would produce temperature changes that would summate because of the passive properties (heat capacity) of the rod. If the two flames were presented in quick succession, a temperature change of 20°C would be produced. If a long time elapses between the presentation of the two flames, there would be little or no summation, because the initial 10°C increase in

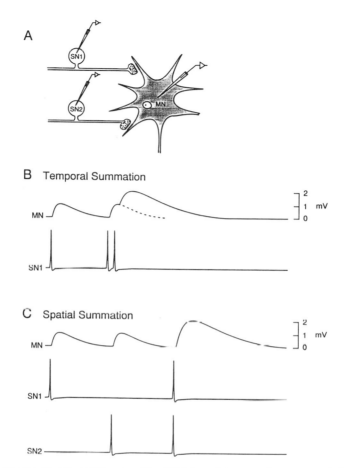

FIG. 13.3. Temporal and spatial summation. **(A)** Intracellular recordings are made from the two sensory neurons (*SN1* and *SN2*) and a motor neuron (*MN*). **(B)** Temporal summation. A single action potential in SN1 produces a 1-mV EPSP in the MN. Two action potentials in quick succession produce a dual-component EPSP, the amplitude of which is approximately 2 mV. **(C)** Spatial summation. Alternative firing of single action potentials in SN1 and SN2 produce 1-mV EPSPs in the MN. Simultaneous action potentials in SN1 and SN2 produce a summated EPSP, the amplitude of which is approximately 2 mV.

temperature would have dissipated. Similarly, potentials in dendrites summate because of the passive properties of the nerve cell membrane. Specifically, the membrane has a capacitance and can store charge. Thus, the membrane will temporarily store the charge of the first EPSP and the charge from the second EPSP will be added to that of the first. The time window for this process of temporal summation is very much dependent on the duration of the postsynaptic potential, however, and temporal summation only occurs if the presynaptic action potentials (and hence, postsynaptic potentials) occur close in time to each other. The time frame is dependent on the passive properties of the postsynaptic membrane, specifically, the time constant.

2. *Spatial Summation.* Spatial summation (Fig. 13.3C), involves a considera-tion of more than one input to a postsynaptic neuron. An action potential in SN1 produces a 1-mV EPSP, just as it did in Fig. 13.3B. Similarly, an action potential in a second sensory neuron (SN2) by itself also produces a 1-mV EPSP. Now, con-sider the consequences of action potentials elicited simultaneously in SN1 and in SN2. The net EPSP is equal to the summation of the amplitudes of the individual EPSPs. Here the EPSP from SN1 is 1 mV, the EPSP from SN2 is 1 mV, and the summated EPSP is 2 mV (Fig. 13.3C). Thus, spatial summation is a mechanism by which synaptic potentials generated at different sites can summate. Spatial summa-tion in nerve cells is determined by the space constant, which is the ability of a potential change produced in one region of a cell to spread passively to other regions of a cell (see Chapter 10).

Again, a thermal analog is useful to help understand this phenomenon. Consider the metal rod as an analogy for a dendrite. As before, temperature is recorded in the middle, but now heat sources can be delivered to either end separately or both ends at the same time. A flame presented to only one end of the metal rod produces a 10°C increase in temperature at the middle of the metal rod; if a flame is presented simultaneously to each end of the metal rod, a 20°C increase in temperature is produced. The temperature changes induced by the two flames presented simul-taneously will summate because of the ability of heat to spread passively along the rod.

In summary, whether a neuron fires in response to synaptic input is dependent, at least in part, on how many action potentials are produced in any one presynaptic excitatory pathway, as well as how many individual convergent excitatory input pathways are activated. The final behavior of the cell is also due to the summation of other kinds of synaptic inputs, specifically, inhibitory synaptic inputs.

Inhibitory Postsynaptic Potentials

Some synaptic events *decrease* the probability of generating spikes in the post-synaptic cell. Potentials produced as a result of these events are called inhibitory postsynaptic potentials (IPSPs). Consider the inhibitory interneuron illustrated in (Fig. 13.2C). Normally, this interneuron would be activated by summating EPSPs from converging afferent fibers. These EPSPs would summate in space and time such that the membrane potential of the interneuron would reach threshold and fire an action potential. This step can be bypassed by artificially depolarizing the inter-neuron to initiate an action potential. The consequences of that action potential from the point of view of flexion motor neurons are illustrated in Fig. 13.2C. The action potential in the interneuron produces a transient increase in the membrane potential of the motor neuron. This transient hyperpolarization (the IPSP) looks very much like the EPSP, but it is reversed in sign.

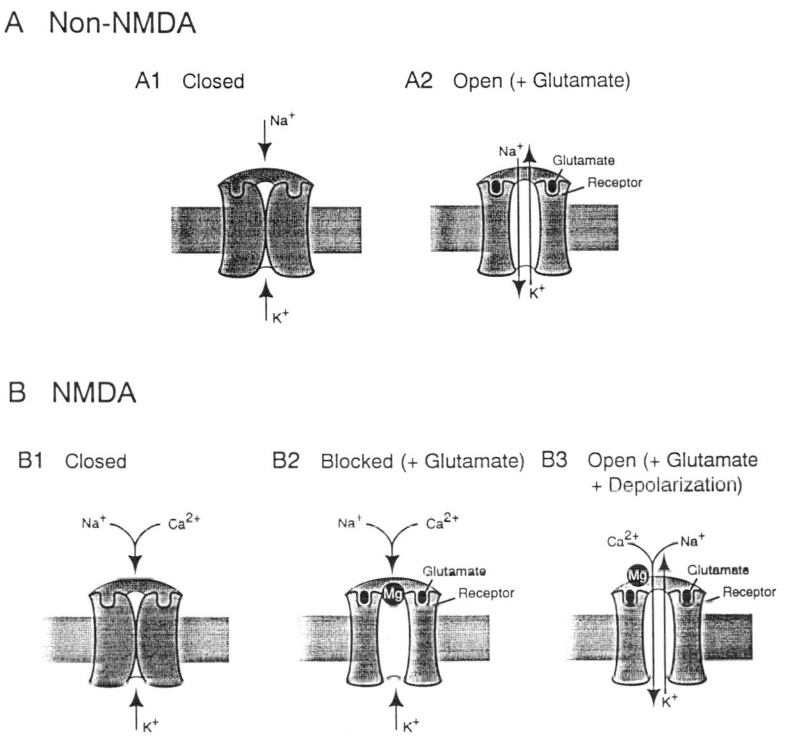

A Non-NMDA

A1 Closed   A2 Open (+ Glutamate)

B NMDA

B1 Closed   B2 Blocked (+ Glutamate) B3 Open (+ Glutamate
                                                          + Depolarization)

FIG. 13.4. Features of non-NMDA and NMDA glutamate receptors. **(A)** Non-NMDA receptors. A1. In the absence of agonist, the channel is closed. A2. Glutamate binding leads to channel opening and an increase in Na^+ and K^+ permeability, which depolarizes the cell to produce an EPSP. **(B)** NMDA. B1. In the absence of agonist, the channel is closed. B2. The presence of agonist leads to a conformational change and channel opening, but no ionic flux occurs because the pore of the channel is blocked by Mg^{2+}. B3. In the presence of depolarization, the Mg^{2+} block is removed and the agonist induced opening of the channel leads to changes in ion flux (including Ca^{2+} influx into the cell).

What are the ionic mechanisms for these fast IPSPs and what is the transmitter substance? Since the membrane potential of the flexor motor neuron is approximately -65 mV, one might expect an increase in the conductance to some ion (or ions) with an equilibrium potential (reversal potential) more negative than -65 mV. One possibility is K^+. Indeed, the K^+ equilibrium potential in spinal motor neurons is approximately -80 mV; so, a transmitter substance that produced a selective increase in K^+ conductance would lead to an IPSP. The K^+-conductance increase would move the membrane potential from -65 mV towards the K^+ equilibrium potential of -80 mV. Although an increase in K^+ conductance mediates IPSPs at some inhibitory synapses, it does not do so at the synapse between the inhibitory interneuron and the spinal motor neuron. At this particular synapse, the IPSP seems to be due to a selective increase in Cl^- conductance. The equilibrium potential for Cl^- in spinal motor neurons is approximately -70 mV. Thus, the transmitter substance released by the inhibitory neuron diffuses across the cleft and interacts with receptor sites on the postsynaptic membrane. These receptors are coupled to a special class of receptor–channels that are normally closed, but when opened they become selectively permeable to Cl^-. As a result of the increase in Cl^- conductance, the membrane potential moves from a resting value of -65 mV toward the Cl^- equilibrium potential of -70 mV.

Like the case of the sensory neuron–spinal motor neuron synapse, the transmitter substance released by the inhibitory interneuron in the spinal cord is an amino acid, but in this case the transmitter is glycine. Indeed, glycine is used frequently in the central nervous system as an inhibitory transmitter, most often in the spinal cord. The most common transmitter associated with inhibitory actions in many areas of the brain is gamma amino butyric acid (GABA). The agents bicuculline and picrotoxin are specific blockers of the actions of GABA. Strychnine is a specific blocker of the actions of glycine.

General Features of Ionotropic-mediated PSPs and the Molecular Biology of the Receptor-Channel Complexes

The general features of ionotropic receptors in the CNS are similar to those of the ligand-gated ACh receptor-channel found in the skeletal muscle. In particular, ionotropic receptors in the CNS for ACh, glycine, and GABA are made up of multiple subunits (usually a pentameric structure) with each of the subunits having four membrane-spanning regions (see Fig. 12.13).

Neuronal nicotinic ACh receptors and GABA receptors have been particularly well characterized. Unlike the skeletal muscle ACh receptor, which is made up of four types of subunits (α, β, γ, δ), the *neuronal* ACh receptor is made up of only two type of subunits (α and β). A considerable diversity of channel properties is possible, however, since there are at least seven different α subunits and three different β subunits. Indeed, it has been estimated that the combinatorial possibilities could result in more than 4,000 different types of receptors. The ionotropic GABA receptor (called $GABA_A$) consists of five different subunits with multiple isoforms. The considerable diversity of $GABA_A$ receptors observed *in vivo* appears due to various combinations of the subunits and their isoforms.

Glutamate is the predominant transmitter with excitatory actions in the CNS, but, until recently, little was known about the structure of its receptor(s). It is becoming clear that glutamate receptors share some features with the ligand-gated channels described above, such as being composed of multiple subunits, each of which has four transmembrane-spanning regions. However, they have a low degree of sequence homology with the subunits of other ionotropic receptors.

Ionotropic glutamate receptors can be divided into two broad classes based on their sensitivity to the agonist N-methyl-D-aspartate (NMDA) and are hence referred to as NMDA and non-NMDA receptors. Both types are located throughout the nervous system, but their relative abundance differs. Non-NMDA glutamate receptors have the functional properties described previously for fast ionotropic-mediated synaptic actions at the sensory neuron–motor neuron synapses (e.g., Fig. 13.2B). Specifically, as a result of glutamate binding, a channel opens that is highly permeable to both Na^+ and K^+ (Fig. 13.4A). The agent 2-amino-3-(3-hydroxy-5-methylisoxazol-4-yl) propanonic acid (AMPA) is a specific agonist of non-NMDA receptors. Inhibitors of non-NMDA receptors include the quinoxaline derivatives 6-cyano-7-nitroquinoxaline-2,3-dione (CNQX) and 6-nitro-7-sulfanoyl-benzo[f] quinoxaline-2,3-dione (NBQX).

NMDA glutamate receptors differ from non-NMDA receptors in four ways. First, they are selectively blocked by the agent 2-amino-5-phosphonovalerate (APV). Second, they have a high permeability to Ca^{2+} as well as Na^+ and K^+. Third, at normal values of the resting potential, the pore of the channel is blocked by Mg^{2+}. Thus, even if glutamate binds to the receptor, there will be no ionic flow (and no EPSP) because the channel is blocked. The block can be relieved by depolarization, which presumably displaces the Mg^{2+} from the pore (Fig. 13.4B). A fourth and final difference between non-NMDA and NMDA channels is a glycine-binding site on the NMDA channel. Glycine must be present for NMDA receptors to function. The physiological role of this binding in regulating the channel is unclear, however. Basal levels of glycine in the vicinity of cells with NMDA receptors seem to be sufficiently high to maintain function of the NMDA receptor. Several of these unique features of the NMDA channel have important physiological consequences. First, activation of the NMDA receptor results in more than a simple EPSP. The influx of Ca^{2+} associated with the channel opening can induce a cascade of biochemical reactions including activation of Ca^{2+}-dependent phosphatases, kinases, and proteases. Second, the dual regulation of the channel by glutamate and voltage (depolarization) allows other synaptic inputs through electrotonic propagation and spatial summation to profoundly regulate the ability of an NMDA-mediated synaptic input to affect a postsynaptic cell. Other examples of such heterosynaptic regulations are described in the following chapter.

Slow Synaptic Potentials Produced by Metabotropic Receptors

A common feature of the types of synaptic actions described above is the direct binding of the transmitter with the receptor-channel complex. An entirely separate class of synaptic actions are due to the indirect coupling of the receptor with the channel. So far, two types of coupling mechanisms have been identified. These include a coupling of the receptor and channel through an intermediate regulatory protein or coupling through a diffusible second messenger system. Receptors that activate the latter mechanism are called *metabotropic*, because they involve changes in the metabolism of second messengers or other compounds such as ATP and phospholipids. Because they are the most common, attention will be focused on the responses mediated by metabotropic receptors.

A comparison of the features of direct, fast ionotropic and indirect, slow metabo-

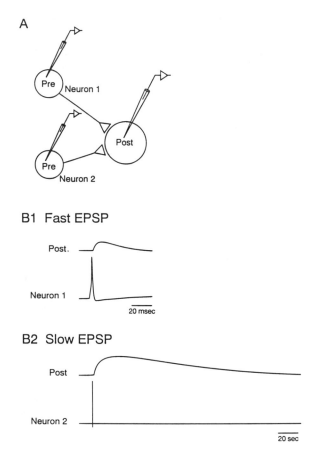

FIG. 13.5. Fast and slow synaptic potentials. **(A)** Two neurons (1 and 2) make synaptic connections with a common postsynaptic follower cell (*Post*). **(B1)** An action potential in Neuron 1 leads to a conventional fast EPSP with a duration of about 30 msec. **(B2)** An action potential in neuron 2 also produces an EPSP in the postsynaptic cell, but the duration of this slow EPSP is more than three orders of magnitude greater than that of the EPSP produced by neuron 1. Note the change in the calibration bar.

tropic-mediated synaptic potentials is shown in Fig. 13.5. Slow synaptic potentials are not observed at every postsynaptic neuron, but Fig. 13.5A illustrates an idealized case in which a postsynaptic neuron receives two inputs, one of which produces a conventional fast EPSP and the other of which produces a slow EPSP. An action potential in neuron 1 leads to an EPSP in the postsynaptic cell whose duration is approximately 30 msec (Fig. 13.5B1). This is the type of potential that might be produced in a spinal motor neuron by an action potential in an afferent fiber. Neuron 2 also produces a postsynaptic potential (Fig. 13.5B2), but its duration (note the calibration bar) is more than three orders of magnitude longer than that of the EPSP produced by neuron 1.

How can a change in the postsynaptic potential of a neuron persist for many minutes as a result of a single action potential in the presynaptic neuron? Possible answers to this question include a prolonged presence of the transmitter due to continuous release, slow degradation, or slow re-uptake of the transmitter; but the mechanism here involves a transmitter-induced change in metabolism of the post-synaptic cell. Figure 13.6 compares the general mechanisms for fast and slow synaptic potentials. Fast synaptic potentials are produced when a transmitter substance binds to a channel and produces a conformational change in the channel causing it to become permeable to one or more ions (both Na^+ and K^+ in Fig. 13.6A). The increase in permeability leads to a depolarization associated with the EPSP (Fig. 13.6A3). The duration of the synaptic event is critically dependent on the amount of time the transmitter substance remains bound to the receptors. The transmitters that have already been mentioned (ACh, glutamate, and glycine) only remain bound for a very short period of time. These transmitters are either removed by diffusion, enzymatic breakdown, or re-uptake into the presynaptic cell. Therefore, the channel closes rapidly.

One mechanism for a slow synaptic potential is shown in Fig. 13.6B. In contrast to the fast PSP for which the receptors are actually part of the ion–channel complex, the channels that produce the slow synaptic potentials are not directly coupled to the transmitter receptors. Rather, the receptors are physically separated and exert their actions indirectly through changes in the metabolism of specific second messenger systems. Figure 13.6B illustrates one type of response that involves the cyclic aden-osine monophosphate (cAMP)–protein kinase A system, but other slow PSPs use other second messenger–kinase systems (e.g., the protein kinase C system). In the case of cAMP-dependent slow synaptic responses, transmitter binding to membrane receptors activates G proteins and stimulates an increase in the synthesis of cAMP. Cyclic AMP then leads to the activation of cAMP-dependent protein kinase [protein kinase A (PKA)], which phosphorylates a channel protein or protein associated with the channel. A conformational change in the channel is produced, which then leads to a change in ionic conductance. Thus, in contrast to a direct conformational change produced by the binding of a receptor to the receptor–channel complex, in this case, a conformational change is produced by protein phosphorylation. Indeed, phosphorylation-dependent channel regulation is one of the general features of slow PSPs.

Another interesting feature of slow synaptic responses is that they are sometimes associated with decreases rather than increases in membrane conductances. For ex-ample, the particular channel illustrated in Fig. 13.6B is selectively permeable to K^+ and is normally open. As a result of the activation of the second messenger, the channel closes and becomes less permeable to K^+. The resultant depolarization may seem paradoxical, but recall that the membrane potential is due to a balance of the resting K^+ and Na^+ permeabilities. The K^+ permeability tends to move the membrane potential toward the K^+ equilibrium potential (-80 mV), whereas the Na^+ permeability tends to move the membrane potential toward the Na^+ equilib-rium potential ($+55$ mV). Normally, the K^+ permeability predominates and the

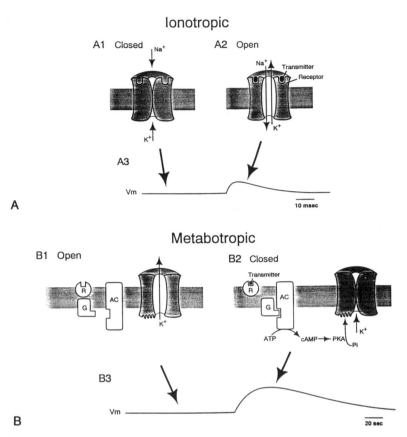

FIG. 13.6. (A) Ionotropic receptors and mechanisms of fast EPSPs. A1: Fast EPSPs are produced by the binding of transmitter to specialized receptors that are directly associated with an ion channel (i.e., a ligand-gated channel). When the receptors are unbound, the channel is closed. A2: Binding of the transmitter to the receptor produces a conformational change in the channel protein such that the channel opens. In this example, the channel opening is associated with a selective increase in the permeability to Na^+ and K^+. The increase in permeability results in the EPSP shown in panel A3. The duration of the EPSPs is directly related to the time that the transmitter remains bound to the receptor. **(B)** Metabotropic receptors and mechanisms of fast slow EPSPs. Unlike fast EPSPs that are due to the binding of a transmitter with a receptor–channel complex, slow EPSPs involve the activation of receptors (metabotropic) that are not directly coupled to the channel. Rather, the coupling takes place through the activation of one of several second-messenger cascades (in this example, the cAMP cascade). B1: A channel that has a selective permeability to K^+ is normally open. B2: Binding of the transmitter to the receptor leads to the activation of a G protein (G) and adenylyl cyclase (AC). The synthesis of cAMP is increased, cAMP-dependent protein kinase (PKA) is activated, and a channel protein is phosphorylated. The phosphorylation leads to closure of the channel and the subsequent depolarization associated with the slow EPSP is shown in panel B3. The response decays are due to both the breakdown of cAMP by cAMP-dependent phosphodiesterase and the removal of phosphate from channel proteins by protein phosphatases (not shown).

resting membrane potential is close to, but not equal to, the K^+ equilibrium potential. If K^+ permeability is decreased because some of the channels close, the membrane potential will be biased toward the Na^+ equilibrium potential and the cell will depolarize.

At least one reason for the long duration of slow PSPs is that second messenger systems are slow (second to minutes). Take the cAMP cascade as an example. Cyclic AMP takes some time to be synthesized, but, more importantly, cAMP levels can remain elevated for a relatively long period of time (minutes) after synthesis. The duration of the elevation of cAMP is dependent upon the actions of cAMP-phosphodiesterase that breaks down cAMP. The duration of an effect could also outlast the duration of the change in the second messenger because of persistent phosphorylation of the substrate protein(s). Phosphate groups are removed from the substrate proteins by protein phosphatases. Thus, the net duration of a response initiated by a metabotropic receptor is dependent upon the actions of not only the synthetic and phosphorylation processes, but also the degradative and dephosphorylation processes.

The activation of a second messenger by a transmitter can have a localized effect on membrane potential through phosphorylation of membrane channels near the site of synthesis. The effects can be more widespread and even longer lasting than depicted in Fig. 13.6, however. For example, second messengers and protein kinases can diffuse and affect more distant membrane channels. Moreover, a long-term effect can be induced in the cell by altering gene expression. For example, PKA can diffuse to the nucleus where it can activate proteins that regulate gene expression.

BIBLIOGRAPHY

Cooper JR, Bloom FE, Roth RH. *The biochemical basis of neuropharmacology*, 6th ed. New York: Oxford University Press, 1991.

Eccles JC. *The understanding of the brain*, 2nd ed. New York: McGraw-Hill, 1977.

Kandel ER, Schwartz JH, Jessell TM. *Principles of neural science*, 3rd ed. New York: Elsevier/North Holland, 1991; Chapters 11 and 12.

Krogsgaard-Larsen P, Hansen JJ, eds. *Excitatory amino acid receptors*. New York: Ellis Horwood Press, 1992.

Shepherd GM, ed. *The synaptic organization of the brain*, 3rd ed. New York: Oxford University Press, 1990; Chapters 1–4, 13.

Siegel GJ, Agranoff BW, Albers RW, Molinoff PB. *Basic neurochemistry*, 5th ed. New York: Raven Press, 1994.

ADDITIONAL READING

Barnard EA. Receptor classes and the transmitter-gated channels. *Trends Biochem Sci* 1992;17:368.

Burke RE, Rudomin P. Spatial neurons and synapses, In: Kandel ER, ed. *Handbook of physiology, Section 1: The nervous system*, vol 1, pt 2. Bethesda: American Physiological Society, 1977: 877–944.

Role LW, Diversity in primary structure and function of neuronal nicotinic acetylcholine receptor channels. *Curr Opin Neurobiol* 1992;2:254.
Spencer WA. The physiology of supraspinal neurons in mammals, In: Kandel ER, ed. *Handbook of
physiology, Section 1: The nervous system*, vol 1, pt 2. Bethesda: American Physiological Society,
1977: 969–1022.

14

Synaptic Plasticity

Up to this point, the process of synaptic transmission has been presented as if it were fairly rigid and stereotyped. Every action potential in a presynaptic cell produced a postsynaptic potential that was identical in its amplitude and duration to previous ones generated by the same cell. This is not generally the case, however. Indeed, one of the interesting and important features of synaptic transmission is that the efficacy (or strength) of the synapse can change profoundly. The process by which the efficacy of the synapse can change is called synaptic plasticity. Synaptic plasticity is important for regulating reflex excitability, the efficacy of synaptic transmission in pain and other afferent pathways, and as a mechanism for learning and memory.

There are two broad categories of synaptic plasticity, namely intrinsic and extrinsic. Intrinsic (also referred to as homosynaptic) refers to a change in synaptic efficacy between two neurons as a result of activity in one or both of the neurons. Extrinsic plasticity (also referred to as heterosynaptic) refers to a change in the efficacy of the synapse between two neurons produced as a result of activity in a third independent neuron or pathway.

One fundamental problem in synaptic plasticity common to both intrinsic and extrinsic types of plasticity is whether the locus of the plasticity is on the pre- or postsynaptic side of the synapse. One possibility is enhanced release of transmitter from the presynaptic terminal. Transmitter release can be considered the product of two factors: the number of vesicles available for release (N) and the probability of release (P). Factors affecting the number of vesicles include synthesis of transmitter, re-uptake of released transmitter, and mobilization of vesicles from storage or reserve pools to release sites in the presynaptic terminal. The probability of release can be affected by any factor that regulates Ca^{2+} influx into the terminal (such as the ability of the spike to invade the terminal or changes in spike height or width), the regulation of levels of intracellular Ca^{2+}, or any other step in the excitation–secretion process. A change in synaptic efficacy could also have a postsynaptic locus. For example, the effectiveness of a fixed amount of transmitter released by a presynaptic cell could be enhanced in several ways: by increasing the number of receptors for the transmitter; by increasing the affinity of the receptor for the transmitter; by increasing the membrane resistance of the postsynaptic cell so that fixed synaptic current produces a greater change in membrane potential; or by changing dendritic morphology so that the postsynaptic current is propagated more effectively to some integrative area of the cell. Changes in morphology of the pre- and/or postsynaptic regions could also result in alterations of the synaptic cleft so that the released transmitter more effectively reaches receptor sites on the postsynaptic cell.

Thus there are many ways to alter synaptic efficacy. Insights into the locus of change for any specific example of synaptic plasticity can be obtained by applying a quantal analysis (see also Fig. 12.18).

QUANTAL ANALYSIS OF SYNAPTIC PLASTICITY

Complete details of the mathematical basis of quantal analysis as well as a discussion of many of the underlying assumptions are beyond the scope of this present chapter. Nevertheless, some of the fundamental concepts can be appreciated by considering how the amplitude and frequency distributions of both the spontaneous miniature synaptic potentials and evoked synaptic potentials would change as a result of either a presynaptic or postsynaptic locus for the modification. Figure 14.1 illustrates distributions for a control synapse (Fig. 14.1A), a synapse whose efficacy is increased by a putative postsynaptic mechanism (Fig. 14.1B), and a synapse whose efficacy is increased by a putative presynaptic mechanism (Fig. 14.1C). Consider first the distributions for the control (unmodified) synapse. Figure 14.1A2 illustrates the mean amplitude of the miniature postsynaptic potentials that are from the spontaneous and random release of single vesicles (quanta) of transmitter from the presynaptic terminal. There is some random variation in the amplitude of each miniature event, but the distribution can be described by a single Gaussian function with a mean of approximately 50 μV (Figure 14.1A2). [Note that the amplitude of potentials produced by quanta in the CNS is considerably less than that at the skeletal neuromuscular junction (see Fig. 12.17).] In contrast to the distribution of spontaneous miniature potentials, the distribution of evoked synaptic potentials is a multimodal function with each peak being an integral multiple of the first (Fig. 14.1A1; see also Fig. 12.18). This distribution is due to the statistical fluctuations in the number of quanta released by each presynaptic action potential. Note also that some action potentials do not produce any synaptic potential. These so-called failures are the extreme case of statistical fluctuations in the release mechanism in which the presynaptic action potential fails to release even a single vesicle.

What are the consequences on these distributions of a postsynaptic mechanism for synaptic plasticity? (One such possibility is enhanced receptor sensitivity.) The *frequency* of spontaneous release of vesicles would be unaffected, but now each vesicle by itself would produce a larger *amplitude* potential in the postsynaptic cell. Hence, the mean value of the distribution of amplitudes of the spontaneous potentials would be shifted to the right (Fig. 14.1B2). Similarly, the peaks on the amplitude distribution of evoked postsynaptic potentials would also be shifted to the right

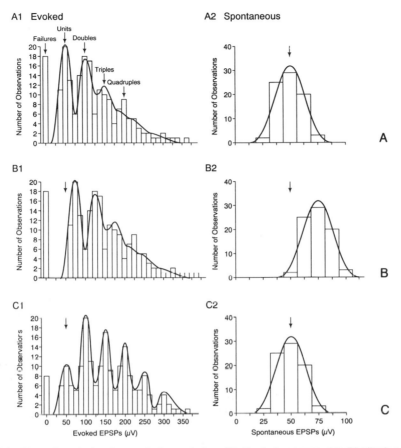

FIG. 14.1. Quantal analysis of synaptic transmission. **(A)** Control synapse. A1: Distribution of the amplitudes of evoked synaptic potentials in a synapse in the central nervous system. There is a multimodal distribution of the amplitudes of postsynaptic potentials evoked by successive action potentials in a presynaptic neuron, which occurs because of statistical fluctuations in the number of quanta released by each action potential. The amplitude of the unitary evoked EPSP is approximately 50 μV and each successive peak is an integral multiple of the first. Some presynaptic action potentials fail to release any quanta. These events are called failures. The frequency of occurrence of the failures is indicated on the left-hand section of the graph. A2: The distribution of spontaneous miniature synaptic potentials can be described by a single Gaussian function with a mean amplitude of approximately 50 μV (*arrow*). **(B)** Changes in quantal parameters produced by an example of synaptic plasticity in which the modification has a postsynaptic locus. B1: The peaks of the amplitude distribution of evoked EPSPs are shifted to the right because now each quantum produces a larger postsynaptic potential. (The position of the arrow corresponds to the location of the unitary peak in A1.) There is no change in the frequency of occurrence of failures, however, nor is there any change in the relative density of the peaks of the distribution of evoked EPSPs. B2: The mean of the distribution of spontaneous miniature postsynaptic potentials is shifted to the right because each quantum produces a larger potential. **(C)** Changes in quantal parameters produced by an example of synaptic plasticity in which the modification has a presynaptic locus. C1: The number of failures decreases because there is an increase in the probability that each action potential in the presynaptic terminal will now release a vesicle of transmitter. Similarly, the probability of releasing multiple vesicles of transmitter with each action potential will be increased. Thus the number of single quanta will decrease and the number of multiple quanta will increase. The density of peaks will shift to the right, but the potentials at which the peaks occur will remain the same as the control synapse. C2: The amplitude distribution of the spontaneous events will be the same as that for the control synapse.

(Fig. 14.1B1). The number of failures would be unchanged, however, since this process is generally a reflection of presynaptic mechanisms (cf. below).

Now consider the consequence on these distributions of a presynaptic mechanism. (One such possibility is enhanced influx of Ca^{2+} through voltage-dependent calcium channels in the presynaptic terminal.) In this case the *amplitude* of any one quantal event will be unchanged. Hence the mean amplitude of the distribution of spontaneous potentials will be unchanged (Fig. 14.1C2). There may also be some increase in the *frequency* of the spontaneous events, but this will not change the distribution of the amplitudes unless, in an extremely rare event, two quanta are released synchronously. The multimodal peaks on the plot of evoked release will distribute around the same peaks of potentials as the control synapse, but there will be a shift to the right in the density of the peaks (Fig. 14.1C1), because the enhanced Ca^{2+} influx will enhance the probability that each action potential will release a greater number of quanta. The number of failures will decrease for similar reasons.

In summary, by performing a quantal analysis before and after a particular manipulation that changes synaptic efficacy, insights into possible loci for the change can be obtained. This information is important by itself, but it is also critical to know to which side of the synapse to direct subsequent biophysical and molecular analyses. Performing a quantal analysis is relatively straightforward. Unfortunately, the results are sometimes difficult to interpret at synapses in the CNS, because the miniature synaptic events and failures are difficult to detect because of their small size, as well as the presence of spontaneous synaptic potentials produced by other presynaptic neurons that converge onto the same postsynaptic cell. Also, some synaptic changes that appear to be postsynaptic in origin (i.e., enhanced receptor sensitivity) can be confused for a subtle presynaptic mechanism (e.g., enhanced filling of vesicles with transmitter). Thus, conclusions regarding the site of plasticity should be based on the application of multiple analyses, of which quantal analysis is but one approach.

INTRINSIC MECHANISMS OF SYNAPTIC PLASTICITY

There are two forms of intrinsic mechanisms. One form known as synaptic depression is associated with an activity-dependent weakening of synaptic strength, whereas the other form known as synaptic facilitation is associated with an activity-dependent enhancement of synaptic strength. It is important to mention that all these plastic changes are not necessarily observed at every synapse. Some synapses exhibit only one form of plasticity, whereas other synapses exhibit multiple forms of both intrinsic and extrinsic plasticity.

Homosynaptic Depression

Figure 14.2 illustrates a spinal motor neuron that receives a synaptic connection from a sensory neuron. Intracellular recordings are made from the motor neuron and

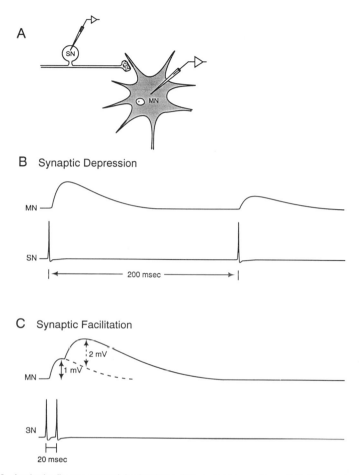

FIG. 14.2. Intrinsic (homosynaptic) plasticity. **(A)** A postsynaptic cell (*MN*) receives a single excitatory input from a presynaptic sensory neuron (*SN*). **(B)** Synaptic depression. The second of two action potentials in the SN produces a smaller EPSP in the MN. **(C)** Synaptic facilitation. The second of two action potentials in the SN produces a larger EPSP in the MN.

the sensory neuron (Fig. 14.2A). Two action potentials are elicited, the second of which occurs about 200 msec after the first (Fig. 14.2B). The second action potential produces an EPSP that is smaller than the first. This decrease in amplitude or strength of the synapse as a result of repeated activation of the presynaptic cell is referred to as homosynaptic depression.

Although many mechanisms could contribute to homosynaptic depression, one that has been implicated repeatedly is a presynaptic mechanism referred to as *depletion*. Specifically, as a result of the release of transmitter by the first action potential, less transmitter will be released by the next action potential. Consequently, the second EPSP will be smaller than the first. This process is depicted graphically in Fig. 14.3A.

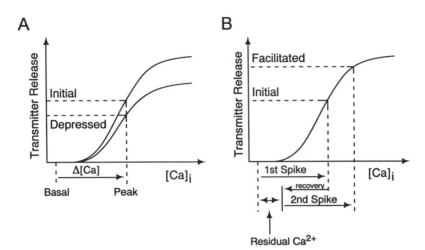

FIG. 14.3. Mechanisms for synaptic depression and facilitation. **(A)** Depletion of the available pool of transmitter is reflected by a downward shift in the sigmoidal relationship between calcium and transmitter release. Thus, even though the second of two action potentials leads to the same Ca^{2+} influx and peak change in intracellular concentration of Ca^{2+} ($[Ca^{2+}]_i$), less transmitter will be released. In this example, it is assumed that transient increases in Ca^{2+} concentration are buffered back to basal (resting) levels during the period between the first and second spike (see below). **(B)** Twin-pulse (paired-pulse) facilitation. The period of time between the two spikes is sufficiently short that the transient increase in Ca^{2+} produced by the first spike has not been buffered completely. Thus, the level of intracellular Ca^{2+} concentration after the second spike is greater than that associated with the first. Because of the sigmoidal relationship, transmitter release is enhanced and the EPSP is facilitated.

As a simple example of depletion, consider a presynaptic terminal that contains a pool of 100 vesicles of transmitter (i.e., $N = 100$) that are available to be released, and that an action potential releases some fixed fraction (maybe 10%) of the available pool (i.e., $P = 0.1$). The initial action potential releases 10 vesicles (10% of 100), and assume that those 10 vesicles produce a potential change of 1 mV in the postsynaptic cell. A second action potential that occurs soon after the first also releases 10% of the available pool of vesicles, but now the number of vesicles available to be released is not 100 but 90. Therefore, the second action potential leads to the release of 9 vesicles, and they produce a smaller EPSP than did the original 10 vesicles. With time, the pool of the transmitter is restored and a second action potential could again release the same 10 vesicles. It might take 500 msec or so to restore the depleted vesicles.

Other mechanisms for homosynaptic depression include desensitization of the postsynaptic receptors and some change in the calcium conductance in the presynaptic terminals. For example, if calcium channels were modulated, the second of two action potentials would lead to less calcium influx, and, even though the same amount of transmitter may be available, less calcium influx would cause less release (i.e., Δ[Ca] in Fig. 14.3A would be less with the second action potential).

Synaptic Facilitation

A phenomenon that has an effect opposite to that of homosynaptic depression is homosynaptic *facilitation*. Figure 14.2C illustrates a particular type of homosynaptic facilitation called paired-pulse or twin-pulse facilitation. With this protocol, the second of two closely initiated EPSPs is greater than the first. This effect is not simply a consequence of temporal summation. Note that the first action potential in the sensory neuron produces a 1-mV EPSP. The second action potential produces a 2-mV EPSP. The summated EPSP from the combined processes of paired-pulse facilitation and temporal summation has an amplitude of 3 mV.

What are the proposed mechanisms for twin-pulse facilitation? Again there are a number of possibilities, but one mechanism that has been implicated repeatedly is a presynaptic mechanism called the "residual calcium hypothesis." To understand this mechanism, it is necessary to consider the nonlinear relationship between calcium concentration and transmitter release at the presynaptic terminal (Fig. 14.3B). As a result of an initial action potential that invades the terminal, there is an increase in the influx of calcium that leads to a transient increase in the concentration of intracellular calcium. The peak Ca^{2+} concentration induces the release of an amount of transmitter sufficient to produce the observed EPSP. The key aspect of the residual calcium hypothesis is that the increase in intracellular Ca^{2+} takes time to be removed by pumps and/or intracellular buffers. If another action potential occurs that leads to a second influx of calcium during the time the calcium concentration is decreasing (but has not yet returned to its basal level), the total peak calcium concentration will be greater, and, according to the nonlinear relationship, more transmitter will be released and a facilitated PSP will be produced.

It may seem paradoxical, but both depression and facilitation can occur at the same synapse. For example, the relationship between Ca^{2+} concentration and transmitter release can compensate for any decrease in the pool of available transmitter. Thus, the facilitatory mechanism (i.e., enhanced P) may override depression (i.e., decreased N). Reconsider the earlier example of a synapse in which an action potential releases 10% of the 100 vesicles of available transmitter. The increased calcium with the second spike can be so much greater that, instead of releasing 10% of the vesicles, 20% of the vesicles could be released. Even if 90 vesicles of the original 100 remain, a greater proportion of them (i.e., 18) would produce a larger EPSP. Facilitation is dependent on the time course of the removal of intracellular Ca^{2+}. No facilitation is observed in Fig. 14.2B, because, during the interspike interval, the calcium levels return to their basal levels.

Another dramatic example of homosynaptic facilitation is called *posttetanic potentiation* (PTP). In Fig. 14.4, a continuous train of action potentials is elicited at a slow rate of approximately once every 5 sec. This rate produces no depression or facilitation. After several control spikes, the presynaptic cell is fired with a high-frequency burst of spikes (the tetanus). Interestingly, when the stimulus rate is returned to its initial level, the EPSPs remain elevated for many seconds to minutes; thus, we have the term posttetanic potentiation. Residual calcium contributes to

PTP, but other presynaptic mechanisms, such as calcium activation of a kinase, may also contribute to aspects of the later phase(s) of the facilitation. A much longer lasting example of this type of potentiation induced by a brief burst of spike activity is called long-term potentiation (LTP) (see below).

EXTRINSIC MECHANISM OF SYNAPTIC PLASTICITY

Extrinsic plasticity refers to changes in the strength of a synapse that are produced by activity in a third independent pathway or neuron. At least one way in which such heterosynaptic effects can be mediated is through a specialized type of synaptic configuration called axo-axonic synapses. Figure 14.5 illustrates these synapses as well as two types of more conventional synaptic arrangements. An axosomatic synapse is a synapse made by an axon of one neuron onto the soma of a postsynaptic neuron, whereas an axodendritic synapse is a synapse made by one axon onto the dendrite of a postsynaptic neuron. An axo-axonic synapse is a synapse made by one axon with the terminal region of another axon. Axo-axonic synapses are a morphological substrate of some examples of extrinsic or heterosynaptic plasticity. Plasticity at axosomatic and axodendritic synapses include, but are not restricted to, homosynaptic depression and facilitation.

Figure 14.6 illustrates two examples of extrinsic modifications. As was the case for intrinsic plasticity, there are two forms of extrinsic modulation; one form in which the synapse is weakened is called presynaptic inhibition, and the other form in which the synapse is strengthened is called presynaptic facilitation.

Presynaptic Inhibition

The basic experimental arrangement to study presynaptic inhibition is illustrated in Fig. 14.6A1. One presynaptic neuron (pre) makes a conventional, synaptic connection and produces a postsynaptic potential in the postsynaptic cell (post). A third modulatory neuron (M1) makes an axo-axonic synapse with the presynaptic neuron. Figure 14.6A2 illustrates the control case in which an action potential in cell pre produced an EPSP in cell post. Also illustrated is the time course of calcium influx or calcium current (I_{Ca}) associated with the action potential-induced opening of voltage-dependent Ca^{2+} channels. Between panels A2 and A3, neuron M1 was fired. Firing neuron M1 by itself produces no changes in the postsynaptic neuron (not shown), but it influences the strength or efficacy of transmission between the presynaptic cell and the postsynaptic cell. Consequently, if transmission between pre and post is retested after firing M1, the amplitude of EPSP in the postsynaptic cell is reduced (Fig. 14.6A3). Thus, as a result of its activation, cell M1 has reduced transmission between two other cells. This phenomenon is called presynaptic inhibition, because the inhibitory effect occurs on the presynaptic side of the synapse. Cell M1 need not have any direct effect on the postsynaptic cell. Rather it somehow modulates the process of synaptic transmission in the presynaptic terminal.

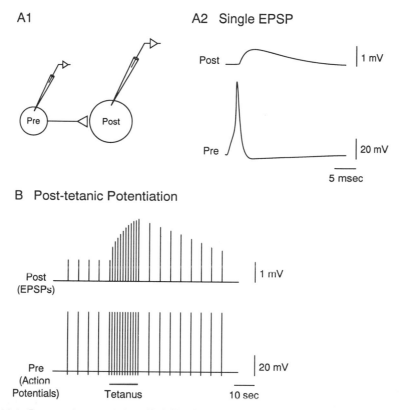

FIG. 14.4. Posttetanic potentiation. **(A1)** Simultaneous intracellular recordings are made from a neuron (*Pre*), which forms a synaptic contact with a second cell (*Post*). **(A2)** An action potential in the presynaptic cell leads to an EPSP in the postsynaptic cell. **(B)** Action potentials are repeatedly elicited in the presynaptic cell at a constant low rate that produces stable EPSPs in the postsynaptic cell. After several stable responses, a brief high-frequency burst of action potentials (tetanus) is elicited in the presynaptic neuron. After the tetanus, the original rate of initiation of action potential is resumed. The EPSPs remain facilitated (potentiated) for several minutes.

Several mechanisms for presynaptic inhibition have been proposed, but the one illustrated here involves a modulation of the presynaptic Ca^{2+} channels by the modulatory neuron. (Putative phosphorylation sites on voltage-gated channels are found on the linker regions between homologous domains; see Chapter 9.) As a result of this modulation, an action potential, elicited after the firing of M1, leads to less calcium influx and, therefore, less transmitter release; the EPSP will be reduced in amplitude. Note that presynaptic inhibition mediated by modulation of Ca^{2+} channels can occur in the absence of any obvious changes in the presynaptic action potential (compare the pre spike in 14.6A2 and A3). This is because the Ca^{2+} conductance is relatively small compared to the larger Na^+ and K^+ conductances that underlie the action potential. Even though the changes in Ca^{2+} current are

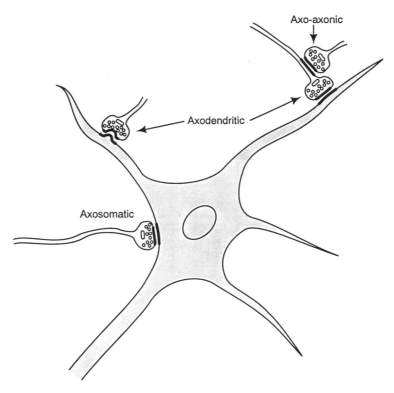

FIG. 14.5. Some basic types of synaptic contacts in the CNS. *Axosomatic:* a presynaptic neuron forms a synaptic contact on the cell body of the postsynaptic neuron. *Axodendritic:* a presynaptic neuron forms a synaptic contact on the dendrite of the postsynaptic neuron. *Axo-axonic:* the presynaptic neuron forms a synaptic contact near the terminal of another neuron.

small relative to the Na^+ and K^+ currents, the change in Ca^{2+} influx has a significant effect on transmitter release. Alternative, although not necessarily mutually exclusive, mechanisms for presynaptic inhibition include modulation of intracellular events associated with mobilization, docking, fusion, and release of vesicles.

Presynaptic Facilitation

The phenomenon of presynaptic facilitation is illustrated in Fig. 14.6B. The morphological arrangement (B1) is similar to that for presynaptic inhibition. Firing the modulatory neuron M2 has no effect by itself in the postsynaptic cell. Rather cell M2 regulates the efficacy of synaptic transmission between the presynaptic (pre) and the postsynaptic (post) cells. The actions of M2 affect the presynaptic side of the synapse, so this phenomenon is called presynaptic facilitation. At least one mechanism for presynaptic facilitation is the regulation of membrane K^+ currents.

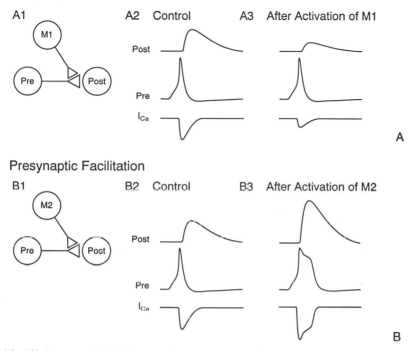

FIG. 14.6. **(A)** Presynaptic inhibition mediated by axo-axonic synapses. A1: The synapse between two pairs of neurons (*Pre* and *Post*) is affected by an axo-axonic synapse from the modulatory neuron M1. A2: An action potential in cell pre produces an EPSP in cell post. The downward deflection on the bottom trace represents the inward calcium current (I_{Ca}) associated with the action potential in the presynaptic neuron. A3: After activation of neuron M1 a subsequent action potential in the presynaptic cell produces an EPSP that is smaller than the control (A2), since activation of M1 leads to a reduction of the voltage-dependent calcium conductance in the presynaptic neuron (*Pre*). With a reduced conductance, the calcium influx (current) is less and the Ca^{2+}-induced release of transmitter is reduced. **(B)** Presynaptic facilitation mediated by axo-axonic synapses. B1: The synapse between two pairs of neurons is affected by an axo-axonic synapse from the modulatory neuron M2. B2: Control EPSP. B3: EPSP produced by an action potential in the presynaptic (*Pre*) neuron after neuron M2 had been activated. The EPSP is larger because the modulatory neuron leads to a reduction of K^+ conductance in the presynaptic (*Pre*) neuron and a subsequent broadening of the action potential and enhanced calcium current (i.e., influx). The enhanced influx of Ca^{2+} leads to enhanced transmitter release and an enhanced EPSP.

After firing cell M2, the subsequent action potential in the presynaptic cell (Fig. 14.6B3) is much broader than it was previously. Unlike the example of presynaptic inhibition, this effect is not due to a direct modulation of Ca^{2+} channels, but rather it is due to a reduction of K^+ conductances. Specifically, as a result of the activation of a second messenger cascade, K^+ channels are phosphorylated resulting in fewer K^+ channels available to be opened and thus a broader action potential. Since the action potential is broader, Ca^{2+} flows into the presynaptic terminal for a greater amount of time and a greater amount of transmitter is released. Note that the Ca^{2+} current is larger in Fig. 14.6B3, not because the modulatory cell has effected the Ca^{2+} channels directly, but rather because the modulatory cell has indirectly influenced the Ca^{2+} channels by affecting K^+ channels. An additional mechanism that contributes to presynaptic facilitation is regulation of events associated with some aspect of the mobilization, docking, or release process. At least some of the specific proteins involved in these processes are regulated by protein phosphorylation.

An implicit assumption in the above discussion is that the modulatory effects of neuron M2 are independent of activity in the presynaptic neuron. In some neurons, a burst of action potentials in the presynaptic neuron shortly before or during the activation of the modulatory neurons leads to an enhancement of the modulatory effects that are above and beyond those produced by the modulatory neuron when activated alone. This type of conjunctive mechanism has been called activity-dependent presynaptic facilitation. It may serve as a neural substrate for some examples of associative learning, such as Pavlovian conditioning.

It is interesting to note that aspects of the mechanisms for synaptic plasticity are recapitulations of aspects of mechanisms for slow PSPs. For example, both the slow EPSP illustrated in Fig. 13.6B and the presynaptic facilitation of Fig. 14.6B are due to a second messenger-induced closure of K^+ channels. Although the mechanisms are similar, the consequences are different. For the case of the slow EPSP, the closure of the K^+ channel leads to a depolarization and an enhancement of the probability of firing a spike in the postsynaptic cell. For the case of presynaptic facilitation, the closure of the K^+ channel is only expressed when the presynaptic cell is probed or "interrogated" by an action potential. Then the consequences are expressed as a broadening of the spike in the presynaptic cell and an enhanced EPSP in the postsynaptic cell.

Long-Term Potentiation

Long-term potentiation (LTP) is a persistent enhancement of synaptic efficacy generally produced as a result of delivering a brief (several second) high-frequency train (tetanus) of electrical stimuli to an afferent pathway. The great difference between the duration of the tetanus and the duration of the subsequent enhancement is the defining characteristic of LTP. Such long-term synaptic enhancement, lasting at least several hours in *in vitro* preparations and weeks in intact preparations, has received growing attention because of the possibility that it is related to natural mechanisms of learning and memory. For example, there have been recent demonstrations of "cooperative" and "associative" influences on LTP, and there appear to be a number of similarities between neural correlates produced by LTP procedures and neural correlates of associative learning (e.g., Pavlovian conditioning). Long-term potentiation has been observed in many regions of the mammalian CNS, in the peripheral nervous system, and in the CNS and neuromuscular junctions of several invertebrates.

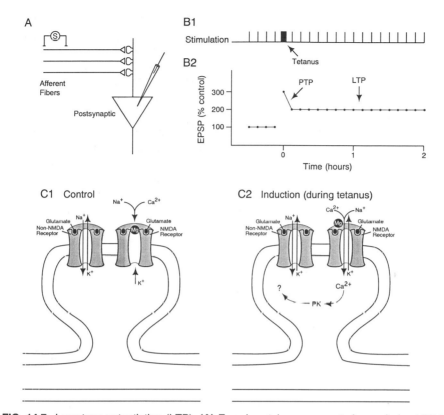

FIG. 14.7. Long-term potentiation (LTP). **(A)** Experimental arrangements for analyzing LTP in the CNS. An intracellular recording is made from a postsynaptic cell (in this case a pyramidal cell in the CA1 region of the hippocampus), and electric shocks are delivered to an afferent pathway (the Schaeffer collateral pathway) that projects to the postsynaptic neuron. **(B)** Stimulus protocol and results. B1: Single weak electric shocks are repeatedly delivered to the afferent pathway. After obtaining several stable baseline responses, a brief high-frequency tetanus is delivered. After the tetanus, the low-frequency test stimulation is resumed. B2: Results. Baseline EPSPs are normalized to their control (pretetanus level). The tetanus produces short-term enhancement (posttetanic potentiation, PTP) followed by an enduring enhancement (LTP) that persists for at least 2 hours. **(C)** Mechanisms for LTP (at the Schaeffer collateral–CA1 pyramidal cell synapse). C1: The spines of the postsynaptic cell have both NMDA and non-NMDA glutamate receptors. Glutamate released by test stimuli activates the non-NMDA receptors. Glutamate released by test stimuli also binds to NMDA receptors but no ions flow through the NMDA channels because they are blocked by Mg^{2+}. C2: The tetanus produces a large postsynaptic depolarization that displaces the Mg^{2+} from the pore of the NMDA channel. Ca^{2+} can now flow into the spine through the NMDA channel and induce a cascade of biochemical reactions (including activation of Ca^{2+}-dependent protein kinases, PK) that lead to a change in synaptic efficacy.

Figure 14.7A illustrates an experimental arrangement for inducing and analyzing LTP. An intracellular recording is made from a postsynaptic neuron that receives monosynaptic excitatory inputs from presynaptic neurons. Brief electric shocks delivered to the afferent pathway lead to the initiation of action potentials in the individual axons in the pathway, and these action potentials propagate to the synaptic terminals. (The rationale for stimulating multiple afferents will become clear when the mechanisms for LTP are discussed below.) The release of transmitter from the multiple afferent terminals produces a summated EPSP in the postsynaptic cell. Test stimuli are repeatedly delivered (Fig. 14.7B1) at a low rate that produces stable EPSPs in the postsynaptic cell (Fig. 14.7B2). After a stable baseline period, a brief high-frequency tetanus is delivered. Subsequent test stimuli produce enhanced EPSPs. The enhancement is associated with at least two temporal domains. There is a large but transient enhancement that represents posttetanic potentiation (PTP) (see above) followed by a stable and enduring enhancement that persists for many hours. This enduring enhancement is referred to as LTP.

Long-term potentiation of the form illustrated in Fig. 14.7B2 has been observed at a number of synapses, but the underlying mechanisms for these different examples of LTP differ. The following discussion focuses on the mechanisms for LTP at a particular synapse, the Schaeffer collateral–CA1 pyramidal cell synapse, in a region of the brain called the hippocampus. The mechanism at this synapse takes advantage of some of the unique properties of the NMDA receptor described in the previous chapter. The Schaeffer collateral axons make synaptic contacts with the pyramidal cells on specialized dendritic structures called spines. The spines have both NMDA and non-NMDA glutamate receptors (Fig. 14.7C). Test stimuli lead to the release of glutamate from the afferent terminal, which diffuses across the synaptic cleft and binds with both types of receptors. Binding to the nonNMDA receptor leads to an increase in the permeability to Na^+ and K^+ and a subsequent small EPSP. Glutamate also binds to the NMDA receptor, but, because of the block by Mg^{2+}, no permeability changes occur, including any to Ca^{2+}. In contrast, the tetanus produces a large depolarization of the spine due to both temporal summation of the effects of the individual EPSPs produced by the non-NMDA receptors at that spine, as well as spatial summation of the effects of EPSPs produced at spines contacted by the other afferent fibers that are conjointly activated by the nerve shock. The resultant large depolarization of the spine displaces Mg^{2+} from the NMDA receptor, which allows Ca^{2+} influx to occur. As described below, the Ca^{2+} influx through the NMDA receptor is essential for the induction of LTP. The rationale for stimulating multiple afferents (Fig. 14.7A) should now be clear. To remove the Mg^{2+} block of the NMDA channel, depolarization from multiple afferents is necessary. This feature of LTP induction is called cooperativity.

The Ca^{2+} influx through the NMDA channel activates one or more Ca^{2+}-dependent protein kinases (PK) (Fig. 14.7C2), and the phosphorylation of substrate proteins leads to an enduring change in synaptic efficacy. The mechanisms for the induction steps subsequent to the activation of kinases are not fully understood. One possibility is a postsynaptic modification, such as an increase in the number of non-NMDA receptors that are available to bind with transmitter released by the posttetanus test stimuli. Alternatively, there may be a retrograde messenger released by the postsynaptic cell to directly affect aspects of the release mechanism in the presynaptic neuron. Quantal analyses to distinguish between these possibilities have yielded conflicting results. It may be that LTP at the Schaeffer collateral– CA1 pyramidal synapse involves modifications of both pre- and postsynaptic processes.

Note that LTP at the synapse illustrated in Fig. 14.7 is a hybrid of homosynaptic and heterosynaptic plasticity. In principle LTP at this synapse could be induced by activation of a single afferent fiber, but the depolarization produced by a single presynaptic neuron (even with a tetanus) is insufficient to relieve the Mg^{2+} block of the postsynaptic NMDA channel. Thus other neurons (i.e., heterosynaptic influences) must be activated as well. Note also that a heterosynaptic type of LTP operationally more similar to the examples of heterosynaptic plasticity described earlier (e.g., Fig. 14.6) could be produced by using two separate inputs to the common postsynaptic cell in Fig. 14.7A. For example, a tetanus to one afferent pathway in conjunction with a weak test stimulus to a separate pathway would lead to an enhancement of the test pathway even though the test pathway was not tetanized. In this case, the tetanized pathway provides the depolarization to relieve the Mg^{2+} block at the synapse of the test pathway. This feature of LTP has been referred to as associativity.

INTEGRATION

Figure 14.8 summarizes the types of events discussed in this and the previous chapter from the perspective of two cells, a presynaptic cell (*pre*) and a postsynaptic cell (*post*), and the ways in which they can be influenced by other inputs. This configuration is idealized, but it is representative of the types of synaptic actions and interactions that are present in the CNS. The presynaptic cell can produce synaptic actions of the classical type in which the released transmitter produces an

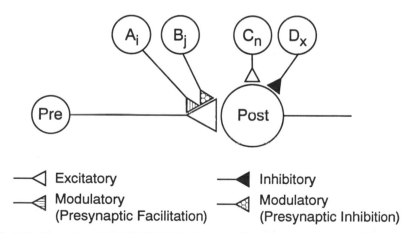

FIG. 14.8. The ability of a presynaptic (*Pre*) cell to activate a postsynaptic (*Post*) cell is dependent on the intrinsic strength of its synaptic connection and changes in its strength due to homosynaptic plasticity. In addition, the synaptic efficacy of the presynaptic neuron can be regulated by classes of presynaptic facilitatory (neurons A_i) and inhibitory (neurons B_j) neurons. Moreover, the ability of the presynaptic neuron to activate the postsynaptic cell will be dependent upon the effects of other excitatory (neurons C_n) and inhibitory (neurons D_x) neurons whose synaptic potentials summate in space and time with those produced by *Pre*.

increased conductance EPSP (or an IPSP). The ability of neuron pre to influence neuron post is dependent on the frequency of firing of cell pre through the process of temporal summation. The ability of cell pre to affect cell post is also dependent upon the intrinsic mechanisms of synaptic plasticity such as homosynaptic depression and facilitation. Moreover, it is dependent upon the ways in which synapse pre is modulated by extrinsic factors, such as presynaptic facilitation and presynaptic inhibition produced by cells A_i and cells B_j. Whether the postsynaptic cell fires is not only dependent upon input from cell pre, but is also dependent upon the activity of cells C_n and cells D_x through the process of spatial summation. Cells C_n are a class of cells that produce EPSPs, and cells D_x are a class of cells that produce IPSPs. At every instance of time, the postsynaptic cell adds the synaptic inputs from the presynaptic cells and "decides" whether to fire an action potential or not. The ability of the postsynaptic cell to process and "evaluate" these multiple inputs through temporal and spatial summation and other mechanisms is called *integration*. If the postsynaptic cell fires, then information is further processed at other postsynaptic neurons.

The nervous system consists of billions of cells, and it has been estimated that an individual neuron can receive up to 10,000 inputs. With billions of cells, each having tens of thousands of plastic synaptic inputs, it is clear why the nervous system is so extraordinarily complex. One of the major challenges of neuroscience is to determine the ways in which specific patterns of cell connectivity and plasticity are used by the various sensory, motor, and higher-order systems of the brain.

BIBLIOGRAPHY

Kandel ER, Schwartz JH, Jessell TM. *Principles of neural science*, 3rd ed. New York: Elsevier/North Holland, 1991; Chapters 13 and 65.

Katz B. *Nerve, muscle, and synapse*. New York: McGraw-Hill, 1966.

Shepherd GM, ed. *The synaptic organization of the brain*, 3rd ed. New York: Oxford University Press, 1990; Chapters 1–4, 11.

Wolpaw JR, Schmidt JT, Vaughan TM, eds. *Activity-driven CNS changes in learning and development*. New York: The New York Academy of Sciences, 1991.

ADDITIONAL READING

Atwood HL, Tse FW. Physiological aspects of presynaptic inhibition. *Adv. Neurosci.* 1993;1:18.

Bliss TVP, Collinridge G. A synaptic model of memory: long-term potentiation in the hippocampus. *Nature* 1993;361:31.

Burke RE, Rudomin P. Spinal neurons and synapses, In: Kandel ER, ed. *Handbook of physiology, Section 1: The nervous system.* vol 1, pt 2. Bethesda: American Physiological Society, 1977:877–944.

Byrne JH. Cellular analysis of associative learning. *Physiol Rev* 1987;67:329–439.

Byrne JH, Zwartjes R, Homayouni R, Critz SD, Eskin A. Roles of second messenger pathways in neuronal plasticity and in learning and memory. In: Shenolikar S, Nairn AC, eds. *Advances in second messenger and phosphoprotein research.* New York: Raven Press, 1993:47–108.

Gingrich KJ, Byrne JH. Simulation of synaptic depression, posttetanic potentiation, and presynaptic facilitation of synaptic potentials from sensory neurons mediating gill-withdrawal reflex in *Aplysia*. *J Neurophysiol* 1985;53:652–669.

Hawkins RD, Abrams TW, Carew TJ, Kandel ER. A cellular mechanism of classical conditioning in *Aplysia*: activity-dependent amplification of presynaptic facilitation. *Science* 1983;219:400–405.

Hawkins RD, Kandel ER, Siegelbaum SA. Learning to modulate transmitter release: themes and variations in synaptic plasticity. *Annu Rev Neurosci* 1993;16:625–665.

Llinas R, Gruner JA, Sugimori M, McGuinness TL, Greengard P. Regulation by synapsin I and Ca^{2+}-calmodulin-dependent protein kinase II of transmitter release in squid giant synapse. *J Physiol (Lond)* 1991;436:257–282.

Llinas R, McGuinness TL, Leonard CS, Sugimori M, Greengard P. Intraterminal injection of synapsin I or calcium/calmodulin-dependent protein kinase II alters neurotransmitter release at the squid synapse. *Proc Natl Acad Sci USA* 1985;82:3035–3039.

Spencer WA. The physiology of supraspinal neurons in mammals, In: Kandel ER, ed. *Handbook of physiology, Section 1: The nervous system*, vol 1, pt 1. Bethesda: American Physiological Society, 1977;969–1022.

Toth PT, Bindokas VP, Bleakman D, Colmers WF, Miller RJ. Mechanism of presynaptic inhibition by neuropeptide Y at sympathetic nerve terminals. *Nature* 1993;364:635–639.

Walters ET, Byrne JH. Associative conditioning of single sensory neurons suggests a cellular mechanism for learning. *Science* 1983;219:405–408.

Zucker RS. Posttetanic potentiation. In: Squire LR, ed. *Encyclopedia of learning and memory*. New York: Macmillan Publishing Company, 1992:527–532.

Appendix

Electrical Circuit Analysis of the Nerve Membrane: The Hodgkin and Huxley Approach

Further understanding of the properties of excitable membranes can be gained by analyzing the equivalent electrical circuit of the membrane, an approach pioneered by Hodgkin and Huxley. In this approach, the membrane is considered to be an electrical circuit composed of a capacitive element in parallel with conductances representing various membrane channels.

PASSIVE PROPERTIES

The equivalent electrical circuit representing the passive properties of an excitable membrane is illustrated in Fig. A.1. The circuit consists of three parallel ele-

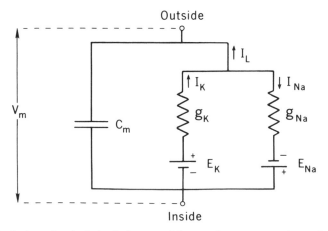

FIG. A.1. Equivalent electrical circuit diagram of the passive nerve membrane (no voltage-dependent conductances). The circuit consists of three parallel branches (see text for details).

ments: a capacitor representing the membrane capacitance; a conductive element (g_K) representing K^+ conductance (permeability) in series with a battery representing the associated K^+ equilibrium potential (E_K); and a conductive element (g_{Na}) representing Na^+ conductance in series with the battery representing the Na^+ equilibrium potential (E_{Na}). In the passive membrane, g_K is normally high and g_{Na} is low. This high resting permeability to K^+ primarily determines the resting potential (see Chapter 8). The Na^+ conductance is low (approximately $\frac{1}{100}$ that of g_K) and reflects the small steady-state leak of Na^+. Because of the leakage currents, the batteries in this circuit would eventually run down. In the cell, however, energy from ATP is used to maintain the ionic concentration gradients and thus keep the batteries "charged."

The ionic current flowing through the two conductances represents the normal leak of K^+ out of the cell and Na^+ into the cell. The current flowing through each branch can be determined by Ohm's law:

$$I = g \times \Delta V \qquad \text{[A.1]}$$

where I is the current created by the movement of ions, g represents the ease with which the ions can flow, and ΔV is the driving force for the movement of ions across the membrane. The potassium current (I_K) is described by

$$I_K = g_K(V_m - E_K) \qquad \text{[A.2]}$$

where V_m is the membrane potential (in this case, the resting potential), E_K is the K^+ equilibrium potential (for the squid giant axon, $E_K = -75$ mV), and ($V_m - E_K$) $- \Delta V$ in Eq. A.1.

The Na^+ current (I_{Na}) is described by

$$I_{Na} = g_{Na}(V_m - E_{Na}) \qquad \text{[A.3]}$$

where E_{Na} is the Na^+ equilibrium potential (for the squid giant axon, $E_{Na} = +55$ mV). The total leakage current (I_L) is simply the sum of $I_K + I_{Na}$, or

$$I_L = I_{Na} + I_K \qquad \text{[A.4]}$$

By substituting Eqs. A.2 and A.3 into Eq. A.4,

$$I_L = g_{Na}(V_m - E_{Na}) + g_K(V_m - E_K)$$
$$= g_{Na}V_m - g_{Na}E_{Na} + g_K V_m - g_K E_K$$
$$= g_{Na}V_m + g_K V_m - g_{Na}E_{Na} - g_K E_K.$$

By rearranging and factoring,

$$I_L = (g_{Na} + g_K) \cdot \left(V_m - \frac{g_{Na}E_{Na} + g_K E_K}{g_{Na} + g_K} \right) \qquad \text{[A.5]}$$

Note that Eq. A.5 is identical in form to Eq. A.1. Therefore, it is possible to simplify the two resistive branches of Fig. A.1 into a single equivalent branch (Fig. A.2) such that the equivalent leakage current is

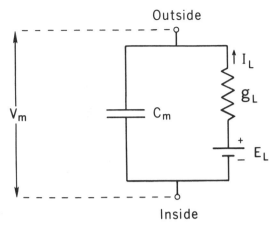

Outside

Inside

FIG. A.2. The circuit of Fig. A.1 can be simplified by combining K^+ conductance (g_K) and its associated equilibrium potential (E_K) with Na^+ conductance (g_{Na}) and its associated equilibrium potential (E_{Na}) into a single leakage conductance (g_L) and battery (E_L).

$$I_L = g_L(V_m - E_L) \qquad\qquad [A.6]$$

where

$$g_L = g_{Na} + g_K \qquad\qquad [A.7]$$

and

$$E_L = \frac{g_{Na}E_{Na} + g_K E_K}{g_{Na} + g_K} \qquad\qquad [A.8]$$

The total equivalent leakage conductance (g_L) is simply the sum of the individual leakage conductances. The equivalent equilibrium potential (E_L) is similarly determined by E_{Na} and E_K but weighted according to the relative values of g_{Na} and g_K.

Equation A.5 can be used to determine the resting potential (V_m) in the steady state. Under those circumstances, no current flows through the membrane capacitance (recall that $I_C = C_m \, (dV/dt)$; in the steady state, $dV/dt = 0$, and thus, $I_C = 0$). Although the conductances of Na^+ and K^+ are different, the net leakage current is zero (the amount of K^+ efflux is equal and opposite to the amount of Na^+ influx). Therefore,

$$I_L = 0 = (g_{Na} + g_K) \cdot \left(V_m - \frac{g_{Na}E_{Na} + g_K E_K}{g_{Na} + g_K} \right)$$

By solving for V_m,

$$V_m = \frac{g_{Na}E_{Na} + g_K E_K}{g_{Na} + g_K} = E_L \qquad\qquad [A.9]$$

Thus, in a passive membrane, the membrane potential is equal to E_L (see also Eq. A.8). Note that when $g_{Na} = 0$, $V_m = E_K$, and when $g_K = 0$, $V_m = E_{Na}$. When g_K is high relative to g_{Na}, the membrane potential will be near the K^+ equilibrium potential. Equation A.9 is the electrical circuit equivalent of the GHK equation described earlier (Chapters 7 and 8). The resting potential is determined by the concentration gradients of K^+ and Na^+ (which determine E_K and E_{Na}) and the relative conductances of the membrane for K^+ and Na^+.

ACTIVE PROPERTIES

The passive membrane model of Fig. A.2 can be expanded to account for the active properties of excitable membranes by including additional parallel branches representing the voltage-dependent Na^+ and K^+ conductance channels. Such an expanded model is illustrated in Fig. A.3. The arrows through the Na^+ and K^+ conductances indicate that they are variable (the conductances change as a function of both voltage and time).

Because current must be conserved, the sum of the currents in all of the four branches must equal zero (Kirchhoff's law). Therefore,

$$0 = C_m \frac{dV}{dt} + I_L + I_{Na} + I_K \qquad [A.10]$$

$$0 = C_m \frac{dV}{dt} + g_L(V_m - E_L) + g_{Na}(V,t)(V_m - E_{Na}) \qquad [A.11]$$

$$+ g_K(V,t)(V_m - E_K)$$

The active conductances, g_{Na} and g_K, are described by $g_{Na}(V,t)$ and $g_K(V,t)$, respectively because g_{Na} and g_K are functions of both voltage and time. The membrane

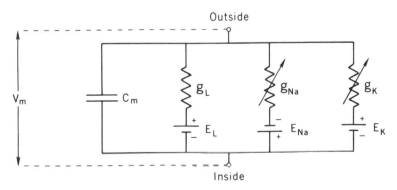

FIG. A.3. Equivalent electrical circuit for a nerve membrane, which includes the voltage-dependent Na^+ conductance (g_{Na}) and voltage-dependent K^+ conductance (g_K) in addition to the leakage conductance (g_L) and membrane capacitance of Fig. **A.2**.

potential at rest or during an action potential can then be found by solving Eq. A.10 for V_m. It is first necessary, however, to determine how g_{Na} and g_K change as functions of voltage and time.

Analysis of the Time and Voltage Dependence of Na$^+$ Conductance

Hodgkin and Huxley modeled Na$^+$ conductance as the product of three factors: maximum Na$^+$ conductance ($\bar{g}_{Na}$); a dimensionless activation term, $m(V, t)$, which varies between 0 at hyperpolarized levels and 1 at large depolarizations; and a dimensionless inactivation term, $h(V, t)$, which varies between 1 at hyperpolarized levels and 0 at large depolarizations. An activation, or m, value of 1 indicates that all of the Na$^+$ channels are open, whereas an m value of 0 indicates that all of the channels are closed. On the other hand, an inactivation, or h, value of 0 indicates that all the channels are inactivated, whereas an h value of 1 indicates that none of the channels are inactivated.

Hodgkin and Huxley determined empirically that Na$^+$ conductance can be described by the equation

$$g_{Na}(V, t) = \bar{g}_{Na} \times m^3(V, t) \times h(V, t) \qquad [A.12]$$

Activation and inactivation do not occur instantaneously with changes in potential, and the m and h terms can be described by first-order differential equations:

$$\tau_m(V) \cdot \frac{dm}{dt}(V,t) + m(V,t) = m(V, \infty) \qquad [A.13]$$

$$\tau_h(V) \cdot \frac{dh}{dt}(V,t) + h(V,t) = h(V, \infty) \qquad [A.14]$$

where $m(V, t)$ and $h(V, t)$ are the instantaneous values of m and h; $m(V, \infty)$ and $h(V, \infty)$ are the steady-state values of the respective activation and inactivation terms at the potential V (i.e., the value that would be achieved at a constant potential); and $\tau_m(V)$ and $\tau_h(V)$ are the respective time constants and represent the rates at which the respective m and h values change when the voltage is changed. Since τ_m and τ_h are also voltage dependent, the rates of activation (and inactivation) are dependent on the membrane potential. Note that with a long, continuous depolarization (i.e., $t = \infty$), $m(V, \infty) = 1$ and $h(V, \infty) = 0$.

Solutions of Eqs. A.12 to A.14 under conditions of a depolarizing step give rise to the voltage clamp results presented earlier. With a constant step depolarization, the explicit solution of Eq. A.13 is a rising exponential, and the explicit solution of Eq. A.14 is a falling or decaying exponential. Since the Na$^+$ conductance is the product of a fast rising exponential and a slower falling exponential, the change in Na$^+$ conductance in response to a step depolarization is of the form of Fig. 9.6.

Solutions of Eqs. A.13 and A.14 require an estimation of the activation and inactivation time constants and the steady-state values of activation and inactivation at

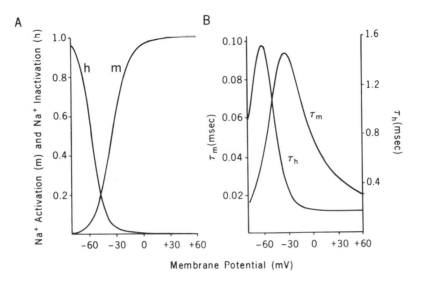

FIG. A.4. Plots of **(A)** the Na$^+$ activation term, m, and Na$^+$ inactivation term, h, and **(B)** the Na$^+$ activation time constant (τ_m) and Na$^+$ inactivation time constant (τ_h) as a function of membrane potential (V_m).

each membrane potential. This information can be obtained from voltage clamp experiments. By stepping the membrane potential of the cell from hyperpolarized levels to a series of fixed depolarizations, both the activation and inactivation time constants as well as the steady-state activation can be estimated from the measured time- and voltage-dependent changes in g_{Na} (see Fig. 9.6). Estimates of the steady-state inactivation parameter, $h(V, \infty)$, can be obtained by clamping from various holding levels to a fixed depolarized level. A plot of the steady-state values of the Na$^+$ activation term (m) and the Na$^+$ inactivation term (h) as a function of voltage is illustrated in Fig. A.4A. Figure A.4B illustrates a plot of the time constant for Na$^+$ activation (τ_m) and the time constant for Na$^+$ inactivation (τ_h) as a function of voltage.

Analysis of the Time and Voltage Dependence of K$^+$ Conductance

Analysis of K$^+$ conductance is similar to that for Na$^+$ conductance but somewhat less complicated, since K$^+$ conductance in the squid giant axon does not inactivate. Hodgkin and Huxley modeled K$^+$ conductance as the product of two factors: the maximum K$^+$ conductance $\bar{g}_K$ and a dimensionless activation term, $n(V,t)$, which varies between 0 at hyperpolarized levels and 1 at large depolarizations. Thus, the n parameter for K$^+$ is analogous to the m parameter for Na$^+$ conductance. They determined that K$^+$ conductance can be described by the equation

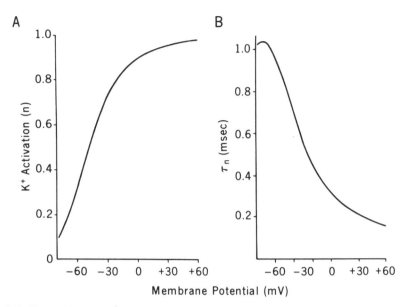

FIG. A.5. Plots of **(A)** the K^+ activation term, n, and **(B)** the activation time constant for the K^+ channel (τ_n) as a function of membrane potential.

$$g_K(V, t) = \bar{g}_K \cdot n^4(V, t) \qquad [A.15]$$

Changes in the n parameter, like changes in m, do not occur instantaneously (indeed, they occur more slowly than m; i.e., τ_n is greater than τ_m) and can also be described by a first-order differential equation,

$$\tau_n(V) \cdot \frac{dn}{dt}(V,t) + n(V,t) = n(V, \infty) \qquad [A.16]$$

where $n(V, t)$ is the instantaneous value of n; $n(V, \infty)$ is the voltage-dependent steady-state value of $n(V, t)$; and $\tau_n(V)$ is the time constant for activation (opening) of the K^+ channel(s). Using the voltage clamp technique (see Fig. 9.11), estimates for $n(V, \infty)$ and $\tau_n(V)$ can be obtained, and Eq. A.16 can be solved for $n(V, t)$. This variable can be substituted into Eq. A.15, and $g_K(V, t)$ can be calculated. A plot of the K^+ activation parameter (n) as a function of voltage is illustrated in Fig. A.5A. In Fig. A.5B, the activation time constant for the K^+ channel (τ_n) is plotted as a function of voltage. Note that the activation time constant for K^+ (τ_n) is an order of magnitude greater than the activation time constant for Na^+ (τ_m). Thus, the opening of K^+ channels is slow compared to the opening of Na^+ channels.

Simulation of the Nerve Action Potential

With estimates of the parameters of Eqs. A.12 to A.16, it is possible not only to simulate the magnitude and time course of Na^+ and K^+ conductances produced by

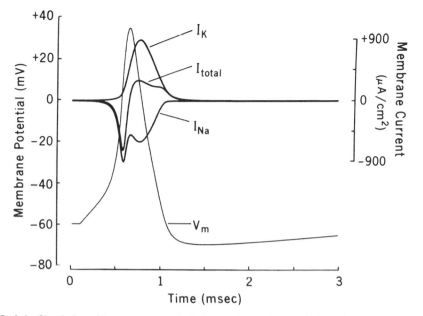

FIG. A.6. Simulation of the action potential in the squid giant axon obtained by solving Eq. A.10; I_{total} is the sum of I_K, I_{Na}, and I_L (see text for details).

a voltage clamp step depolarization, but also to reconstruct or simulate the action potential itself. To do this, Eq. A.10 must be solved for V. When voltage (membrane potential) changes with time, there is no explicit solution for Eq. A.10. It can be readily solved, however, on a digital computer using numerical integration techniques.

To simulate an action potential, an extrinsic suprathreshold stimulus must be applied. This can be accomplished by adding a constant depolarizing current to Eq. A.10. Solution of Eq. A.10 yields a simulated action potential that is essentially identical to the one recorded experimentally. The results of such a simulation are illustrated in Fig. A.6. Also plotted are the individual currents (I_{Na} and I_K) underlying the initiation and repolarization of the action potential. Note the inflection in I_{Na} at the peak of the action potential. This is due to the fact that as the peak of the action potential approaches E_{Na} ($+55$ mV), the driving force for I_{Na} (i.e., $V_m - E_{Na}$), is reduced, and I_{Na} becomes smaller, despite the large Na$^+$ conductance (g_{Na}) (compare Fig. A.6 with Fig. 9.12).

NOVEL IONIC CONDUCTANCE MECHANISMS CONTRIBUTING TO THE INTEGRATIVE ACTION AND FIRING BEHAVIOR OF NEURONS

Investigations of the squid giant axon, as summarized above, have provided significant insights into the mechanisms of neuronal excitability and conduction. It is becoming increasingly clear, however, that neurons often possess additional types of membrane conductances and that this ensemble of membrane conductance endows neurons with a rich repertoire of electrical properties. For example, some cells fire spontaneously in the absence of synaptic input; others fire in bursts; and, still others are silent. Their responses to applied currents also differ. Some cells give long trains of action potentials to depolarizing currents. Others give only one or two initial spikes, and yet others only initiate spikes after a delay period. The diverse biophysical properties of neurons can also play a significant role in determining how a neuron will respond to synaptic input. This section describes selected examples of some of the novel ionic conductance mechanisms that contribute to specific firing properties of individual neurons.

Fast Transient K$^+$ Current (I_A)

The fast transient K$^+$ current or A current contributes to repetitive firing and delayed firing patterns in neurons. This K$^+$ current differs from the classical delayed K$^+$ current (see above) in at least four fundamental ways. First, and as implied by its name, its activation kinetics are more rapid than the activation kinetics of the delayed K$^+$ current. Second, the current exhibits pronounced inactivation (not unlike that exhibited by the Na$^+$ current). Third, the current is generally activated with depolarizations less than those necessary to activate significantly the Na$^+$ or delayed K$^+$ currents. Fourth, the current is inactivated at levels of membrane potential near the resting potential of some neurons (e.g., -50 mV). This latter property has a profound effect on the contribution that this current makes to the firing pattern of different neurons. For example, a depolarizing stimulus would result in little, if any, current flow through A channels if the resting potential of a

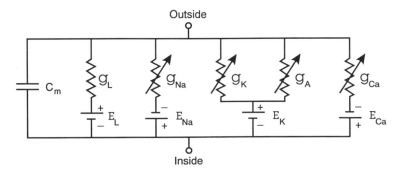

FIG. A.7. Equivalent electrical circuit of a neuron with an A current. The equivalent circuit consists of six membrane components including: (1) a membrane capacitance (C_m); (2) a leakage conductance (g_L) and its associated equilibrium potential (E_L); (3) a Na$^+$ conductance (g_{Na}) with its associated equilibrium potential (E_{Na}); (4) a K$^+$ conductance (g_K) with slow kinetics and its associated equilibrium potential; (5) a K$^+$ conductance channel (g_A) having rapid kinetics and its associated equilibrium potential (E_K); and, (6) a Ca^{2+} conductance (g_{Ca}) with its associated equilibrium potential (E_{Ca}).

neuron was approximately -50 mV, because at this potential most, if not all, of the A channels would be inactivated already. In contrast, the consequences of A channels would be expressed if the resting potential was sufficiently negative (e.g., -70 mV) so that the inactivation of the A channels was removed. Consequently, a depolarizing stimulus could activate the A channels, and current through these channels would affect the membrane potential in a time- and voltage-dependent manner.

Figure A.7 illustrates an equivalent electrical circuit for a nerve cell membrane that includes a conductance (g_A) representing A channels. The circuit also includes conductance branches representing leakage channels (g_L), voltage-dependent Na$^+$ channels (g_{Na}), voltage-dependent delayed K$^+$ channels (g_K), and voltage-dependent calcium channels (g_{Ca}). The A conductance can be described by equations similar to those used for the voltage-dependent Na$^+$ conductance (i.e., A.12, A.13, and A.14), although the specific values of the parameters for the maximum conductance and steady-state activation and inactivation would obviously differ.

The response of a neuron with an A current to a 2.5-sec depolarizing current pulse is illustrated in Figure A.8. For this cell, the leakage conductance and its associated equilibrium potential result in a resting potential of approximately -75 mV. The current pulse leads to an initial relatively large depolarization, but no initial action potentials are triggered as might be expected. Rather, the initial depolarization is followed by a several-second period, during which time the membrane potential depolarizes slowly. Eventually, threshold is reached and a train of action potentials is produced. The Na$^+$ current (I_{Na}) and, to a smaller extent, Ca^{2+} current (I_{Ca}) underlie the initiation of the action potential, whereas the delayed K$^+$ current (I_K) and fast K$^+$ current (I_A) contribute to the spike repolarization. The fast transient K$^+$ current, however, also appears to play a critical role in mediating the delayed firing behavior. The high resting potential removes the inactivation of the fast outward current. Thus, the stimulus-induced depolarization activates this current maximally, and its activation opposes the effectiveness of the depolarizing current pulse to drive the cell to threshold. But, with time, the fast transient current inactivates, and the current pulse becomes more effective in depolarizing the cell. After a several-second rise, the membrane potential reaches threshold and a train of action potentials is elicited.

In summary, the fast transient K$^+$ current can play a unique role in neuronal information processing. It provides a simple mechanism by which delays can be built into neural circuits. Also, it can be considered to act as a braking mechanism ensuring that only strong and/or long-lasting stimuli initiate firing in a neuron. Note also that the relative contribution of the A current can be regulated by any mechanism that affects the resting potential (e.g., ionic pumps and exchangers, fast and slow synaptic input). Thus, a secondary consequence of an inhibitory synaptic input to a neuron with a relatively low resting potential (i.e., -50 mV) might be a hyperpolarization of the cell sufficient to remove the steady-state inactivation of the A channel. A subsequent depolarizing stimulus during the period of synaptically induced hyperpolarization would then be less likely to fire the cell, because the cell is

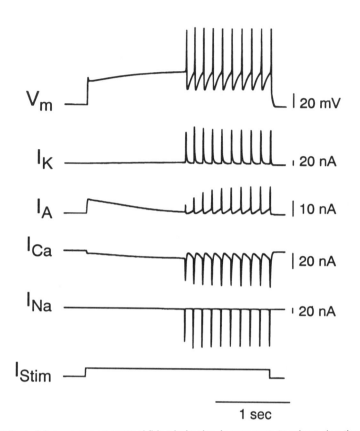

FIG. A.8. Effect of A current on neuronal firing behavior. In response to a long-duration current pulse, there is an initial rapid depolarization followed by a slower increase in the level of depolarization. After several seconds, the depolarization reaches threshold and initiates a burst of action potentials. The activation and inactivation sequence of the fast K^+ current (I_A) appears to underlie the selective response of the cell to long-duration stimuli. (Modified from J. H. Byrne, *J Neurophysiol* 1980;43:652–668.)

further from threshold *and also* because the depolarizing stimulus would activate an opposing A current.

Calcium-Inactivated Ca^{2+} Current and Its Role in Bursting Activity

Some neurons exhibit spontaneous bursts of action potentials separated by quiescent periods (Fig. A.9). These bursts can occur in the absence of any synaptic input. The ionic currents responsible for such endogenous bursting behavior appear to differ among various bursting neurons, but one particularly well-described current is a low-threshold Ca^{2+} current called the slow inward current (I_{SI}). This current differs from those discussed previously, because it is not inactivated by voltage but rather by the level of intracellular Ca^{2+}.

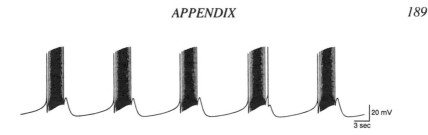

FIG. A.9. Endogenous bursting activity in the R15 neuron of *Aplysia*. In the absence of any synaptic input, the cell generates a burst of spikes at a rate of approximately once every 10–15 seconds. Within individual bursts, the rate of frequency of action potentials is associated with an accelerating phase, the maximum frequency of which occurs about half-way through the burst; thereafter, it decelerates. After the last spike, there is a characteristic spike after depolarization, which is followed by a prominent and deep postburst hyperpolarization. The hyperpolarization relaxes and the cycle repeats itself. (From D. A. Baxter and J. H. Byrne, unpublished observations.)

Figure A.10 illustrates an equivalent electrical circuit for the membrane of a well-described invertebrate bursting neuron. The circuit consists of the parallel combination of a membrane capacitance (C_m), a leakage conductance (g_L), a voltage-dependent Na$^+$ conductance (g_{Na}), and a delayed K$^+$ conductance (g_K), all similar to those described previously for the squid giant axon. The circuit also includes a voltage-dependent Ca^{2+} conductance (g_{Ca}) with basic properties similar to those of g_{Na}, albeit with different kinetics and voltage sensitivity. Of key importance to the bursting activity is the conductance labeled g_{SI}. This is a Ca^{2+} channel that has a

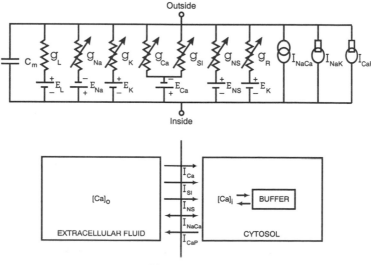

FIG. A.10. Equivalent electrical circuit and calcium balance for a bursting neuron. See text for details. (Modified from C. Canavier, J. W. Clark, and J. H. Byrne, *J Neurophysiol* 1991;66:2107–2124.)

low threshold for activation and is *inactivated* by elevated intracellular levels of Ca^{2+}. Another novel channel in this neuron is nonspecific inward conductance (g_{NS}). This channel appears to be permeable to Na^+, Ca^{2+}, and K^+ and is *activated* by elevated intracellular levels of Ca^{2+}. Finally, the circuit also includes an inward rectifier conductance (g_R). This is a voltage-dependent K^+ conductance, but with a voltage-dependence opposite to those described previously. Specifically, the conductance of this channel increases in response to membrane hyperpolarization. Because at least two of the membrane conductances (i.e., g_{SI} and g_{NS}) are sensitive to the levels of intracellular ions, the model of this neuron must also include descriptions of the various ionic pumps, exchanger mechanisms, and intracellular regulatory mechanisms. Hence, the equivalent circuit includes a Na^+–Ca^{2+} exchanger

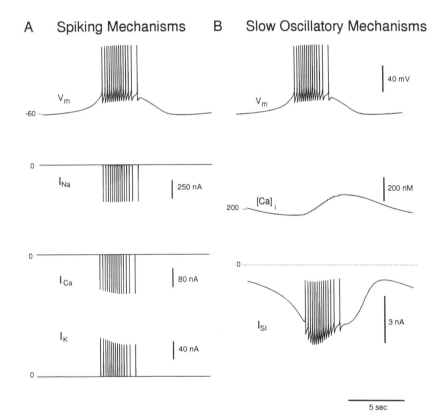

FIG. A.11. Membrane currents contributing to the electrical activity of a bursting neuron. **(A)** Spiking mechanisms. The *top panel* illustrates a burst of spikes in the neuron. The spikes are preceded by a slow depolarization of the membrane potential (V_m) and followed by a postburst hyperpolarization. The large-amplitude currents contributing to the spikes during a burst are plotted below the membrane potential. **(B)** Mechanisms contributing to the slow oscillations. The burst in the top panel of A is repeated in the *top panel* of B. The *second trace* illustrates the corresponding levels of intracellular calcium ([Ca]$_i$), and the *bottom trace* illustrates the slow inward current (I_{SI}) that primarily contributes to the slow oscillations of potential. (Modified from C. Canavier, J. W. Clark, and J. H. Byrne, *J Neurophysiol* 1991;66:2107–2124.)

current (I_{NaCa}), a Na$^+$–K$^+$ pump current (I_{NaK}) and a Ca^{2+} pump current (I_{CaP}). The model also contains material balance equations that describe the buffering and regulation of intracellular Ca^{2+}.

The results of a simulation of the model are shown in Fig. A.11. The model provides the following explanation of bursting activity. Essentially, the slow inward current I_{SI} together with the mechanisms for Ca^{2+} regulation are responsible for the slow cyclic variation in membrane potential. The slow inward current (I_{SI}) is activated at relatively hyperpolarized membrane potentials. Thus, even at the initial membrane potential of −60 mV this inward current slowly depolarizes the membrane potential (Fig. A.11B). When the potential is more positive than the spike threshold for the cell, the cell discharges a burst of action potentials. Action potentials are mediated by the conventional voltage-dependent currents (i.e., I_{Na}, I_{Ca}, I_K) (Fig. A.11A). The calcium influx during the burst leads to an increase in the intracellular levels of Ca^{2+}, which in turn leads to an inactivation of I_{SI} (Fig. A.11B). The slow oscillation begins to wane and the burst is terminated. During the interburst hyperpolarization, Ca^{2+} is removed from the cell (Fig. A.11B), and so the Ca^{2+}-dependent inactivation of I_{SI} is gradually removed. I_{SI} increases; the cell depolarizes; and, the cycle repeats itself.

BIBLIOGRAPHY

Adams WB, Levitan I. Voltage and ion dependences of the slow currents which mediate bursting in *Aplysia* neurone R15. *J Physiol (Lond)* 1985;360:69–93.

Baxter DA, Byrne JH. Ionic conductance mechanisms contributing to the electrophysiological properties of neurons. *Curr Opin Neurobiol* 1991;1:105–112.

Byrne JH. Analysis of ionic conductance mechanisms in motor cells mediating inking behavior in *Aplysia california*. *J Neurophysiol* 1980;43:630–650.

Byrne JH. Quantitative aspects of ionic conductance mechanisms contributing to firing pattern of motor cells mediating inking behavior in *Aplysia california*. *J Neurophysiol* 1980;43:651–668.

Canavier CC, Clark JW, Byrne JH. Simulation of the bursting activity of neuron R15 in *Aplysia*: Role of ionic currents, calcium balance, and modulatory transmitters. *J Neurophysiol* 1991;66:2107–2124.

Connor JA, Stevens CF. Voltage clamp studies of a transient outward membrane current in gastropod neural somata. *J Physiol Lond* 1971;243:21–30.

Hille B. Ionic basis of resting and action potentials, In: Kandel ER, ed. *Handbook of physiology, Section 1: The nervous system*, vol 1, Pt 1. Bethesda; American Physiological Society, 1977:99–136.

Hodgkin AL, Huxley AF. A quantitative description of membrane current and its application to conduction and excitation in nerve. *J Physiol* 1952;117:500–544.

Kramer RH, Zucker RS. Ca^{2+} regulated currents in bursting neurones. *J Physiol (Lond)* 1985;362:131–160.

McAllister RE. Two programs for computation of action potentials, stimulus responses, voltage-clamp currents, and current-voltage relations of excitable membranes. *Comp Progr Biomed* 1970;1:146–166.

Moore JW, Ramon F. On numerical integration of the Hodgkin and Huxley equations for a membrane action potential. *J Theor Biol* 1974;45:249–273.

Noble D. Applications of Hodgkin-Huxley equations to excitable tissues. *Physiol Rev* 1966;46:1–50.

Partridge LD, Stevens CF. A mechanism for spike frequency adaptation. *J Physiol (Lond)* 1976;256:315–332.

Shepherd GM, ed. *The synaptic organization of the brain*, 3rd ed. New York: Oxford University Press, 1990; Chapters 2 and 13.

Yamada WM, Koch C, Adams PR. Multiple channels and calcium dynamics. In: Koch C, Segev I, eds., *Methods in neuronal modeling: from synapses to networks*, Cambridge: MIT Press, 1989:97–133.

Ziv I, Baxter DA, Byrne JH. Simulator for neural networks and action potentials: description and application. *J Neurophysiol* 1994;71:294–308.

Subject Index

A
A current, neuron, 186–188

Absolute refractory period, action potential, 100–102

Accommodation, action potential, 102

Acetylcholine
  iontophoresis, 127–129
  K^+ permeability, 130
  Na^+ permeability, 130
  role, 127, 128

Acetylcholine receptor, structure of ligand-gated, 132

Acetylcholine release
  end-plate potential, 137–141
  miniature end-plate potential, 137–141
  quantal nature, 136–141

Action potential
  absolute refractory period, 100–102
  accommodation, 102
  all-or-nothing, 71, 75
  duration, 71
  frequency, 71
  Goldman–Hodgkin–Katz equation, 78–79, 83–84
  ion channel
    specificity, 96–99
    tetraethylammonium, 96–99
    tetrodotoxin, 96–99
  ionic mechanisms, 80–102
  K^+ concentration, 99
  Na^+ concentration, 99
  nerve
    extracellular recording, 70–71
    intracellular recording, 73–75
    K^+ conductance, 91–95
    Na^+ conductance, 91–95
    simulation, 184–185
    sodium hypothesis, 80–83
  optic nerve, 71

propagation, 104–111
  myelinated axon conduction, 110–111
  principles, 104–105
  rate, 105–111
  sequence of steps, 105
  space constant, 107–110
  time constant, 105–107, 109–110
  velocity, 105–111

relative refractory period, 100–102

repolarization, voltage-dependent K^+ conductance, 91–95

skeletal neuromuscular junction, 142

threshold, 100

undershoot, 75

Active transport, carrier-mediated transport, 50–54
  primary, 51–52
  secondary, 52–54

Adenosine triphosphatase, carrier, 51–52

Air-water interface, phospholipid, 2

Amphiphatic molecule, 4

Aqueous pathway
  ion diffusion, 37
  lipid bilayer, 37

Axon, extracellular recording from single, 113–115

B
Bernstein's hypothesis, resting potential, 75–76
  testing, 75–78

Biological channel, ion diffusion, 37–44

Biological membrane
  hydrophilic molecule, larger, 45
  multivalent ion, 45
  osmotic flow, defined, 15
  permeability coefficient, 11–12
  water flow, pathways, 21–24

Bursting activity, calcium-inactivated Ca^{2+} current, 188–191